P9-CFQ-219

Precalculus Mathematics: Functions and Graphs

HARPERCOLLINS COLLEGE OUTLINE

Precalculus Mathematics: Functions and Graphs

Lawrence A. Trivieri, M.A.
DeKalb College

HarperPerennial
A Division of HarperCollinsPublishers

PRECALCULUS MATHEMATICS: FUNCTIONS AND GRAPHS. Copyright © 1993 by HarperCollins Publishers, Inc. All rights reserved. Printed in the United States of America. No part of this book may be used or reproduced in any manner whatsoever without written permission except in the case of brief quotations embodied in critical articles and reviews. For information address HarperCollins Publishers, Inc., 10 East 53rd Street, New York, NY 10022.

An American BookWorks Corporation Production
Project Manager: Mary Mooney
Editor: Gloria Langer, Ph.D.

Library of Congress Cataloging-in-Publication Data

Trivieri, Lawrence A.
    Precalculus mathematics / Lawrence A. Trivieri.
        p.   cm.   —   (HarperCollins college outline)

    Includes index.
    ISBN: 0-06-467165-8
    1. Functions.   2. Algebra—Graphic methods.   I. Title
II. Series.
QA331.3.T76   1993
515—dc20   92-53295

93   94   95   96   97   ABW/RRD   10   9   8   7   6   5   4   3   2   1

Contents

Preface

This text can be used in one of two ways. If you are enrolled in a course, it can be used to supplement the text or materials that you are currently using; it can also be used as a textbook.

If you are using it as a supplement, you can start at the beginning and work your way through the text in sequence; or you can determine which topics you wish to study, and in which order. In each section of the text, there are many solved exercises. Study as many of them as you think are necessary to understand the topic well. If you have difficulty with a particular topic, study the brief text material given in the section. Examine carefully the first couple of solved exercises until you understand what is being discussed. For the balance of the exercises, write down the statement of the exercise on a sheet of paper. Solve the exercise on your own. Then compare your solution with the one given in the text. Do not move on to the next topic until you completely understand the what you are working on. Don't hesitate to ask for additional help from your instructor or from the mathematics lab at your college, if there is one.

If you have been away from the study of mathematics for some time and are thinking about returning to, or starting, college, this book will help you to review some basic precalculus mathematics prior to enrolling in a calculus course. A graphics calculator is recommended for use with this text.

I wish you well in your studies. Any comments or suggestions that you care to offer can be made to me through the publisher, HarperCollins Publishers, Inc.

Several people deserve my gratitude for their assistance in the completion of this project. Fred N. Grayson asked me to write the manuscript. Sincere thanks for their exemplary work go to Mary Mooney, project editor; Gloria Langer, reviewer; Alexander Kopelman, copyeditor; and Robert Bell, accuracy checker.

Finally, I wish to thank my wife, Joyce B. Trivieri, for her loving support, encouragement, and patience throughout this project.

This study guide is dedicated to the loving memory of my father, Anthony Trivieri.

Lawrence A. Trivieri

Precalculus Mathematics: Functions and Graphs

1

The Real Numbers and Their Properties

In this chapter, we review the real numbers and their various subsets, as well as properties associated with these numbers. Ordering of real numbers, inequalities, and absolute values are also reviewed.

1.1 THE REAL NUMBERS

In this section, we review the various subsets of the real numbers together with some of their properties.

The Natural Numbers

The **natural numbers**, denoted by the symbol N, are also called the counting numbers. We have

$$N = \{1, 2, 3, 4, 5, 6, 7, ...\}$$

There are infinitely many natural numbers. The smallest natural number is 1; there is no largest natural number.

The Whole Numbers

Notice that 0 is not a natural number. If we include 0 with all of the natural numbers, we have the set of **whole numbers**, denoted by the symbol W. We have

$$W = \{0, 1, 2, 3, 4, 5, 6, 7, ...\}$$

There are infinitely many whole numbers. The smallest whole number is 0; there is no largest whole number.

The Integers

Each natural number has an opposite. For instance, the opposite of the natural number 3 is –3, the opposite of the natural number 12 is –12, and so forth. If we include all of the natural numbers, all of their opposites, and the whole number 0, we have the set of **integers**, denoted by the symbol Z. We have

$$Z = \{..., -5, -4, -3, -2, -1, 0, 1, 2, 3, 4, 5, ...\}$$

There are infinitely many integers. There is no smallest integer nor largest integer. The set Z can be subdivided into three disjoint sets:

- $\{1, 2, 3, 4, 5, ...\}$ is the set of all **positive** integers.
- $\{..., -5, -4, -3, -2, -1\}$ is the set of all **negative** integers.
- $\{0\}$ is the set containing the integer **zero**.

The Rational Numbers

The set of rational numbers, denoted by the symbol Q, is the set of all numbers that can be expressed as the quotient of two integers, provided that the denominator is not equal to 0. We have

$$Q = \{ \frac{a}{b} \mid a \text{ and } b \text{ are integers, } b \neq 0 \}$$

Some examples of rational numbers are: $\frac{-2}{3}, \frac{4}{7}, \frac{1}{-4}, \frac{-7}{-8}, \frac{23}{-7}$, and $\frac{-57}{101}$. Note that all integers are also rational numbers. For instance, the integer 3 can be written as $\frac{3}{1}$; the integer –9 can be written as $\frac{-9}{1}$; and so forth. Both repeating and terminating decimals are also rational numbers. For instance, 23.78, $0.1\overline{6}$, $2.\overline{17}$, and 0.1234 are rational numbers.

Irrational Numbers

Not every decimal number terminates or repeats. The set of all numbers that have decimal representations that do not terminate or repeat is called the set of **irrational numbers**, denoted by the symbol J. Some examples of irrational numbers are $\sqrt{3}, -\sqrt{8}, \pi, \frac{\pi}{4}$, and $4 - 2\sqrt{11}$.

The symbol π represents the ratio of the circumference of a circle to its diameter. The number π can be approximated as 3.1416 to four decimal places. Note that 3.1416 is a rational approximation for the irrational number π.

The Real Numbers   The set of **real numbers**, denoted by the symbol R, is the set of all numbers that are either rational or irrational. All examples of numbers given above are examples of real numbers. However, note that not all numbers are real numbers. Recall from your algebra courses that the solutions of some quadratic equations are complex numbers. These will be considered briefly later in the text.

EXERCISE 1.1

For each of the following, determine to which of the sets N, W, Z, Q, J, or R the number belongs. If none, so state.

a) -7

b) $\sqrt{13}$

c) 2.45

d) $\dfrac{-7}{12}$

e) 7π

f) $\sqrt{9}$

g) $\sqrt{-7}$

h) $\sqrt{2.25}$

i) 0

j) -1.0203

SOLUTION 1.1

a) Z, Q, R

b) J, R

c) Q, R

d) Q, R

e) J, R

f) Since $\sqrt{9} = 3$, we have N, W, Z, Q, R

g) None

h) Q, R

i) W, Z, Q, R

j) Q, R

EXERCISE 1.2

What, if any, is

a) the least of the non-negative integers; and
b) the greatest of the nonpositive integers?

SOLUTION 1.2

a) Non-negative means not negative. Hence, the least of the non-negative integers is 0.
b) Nonpositive means not positive. Hence, the greatest of the nonpositive integers is 0.

1.2 ORDERING OF THE REAL NUMBERS

One of the most important properties of the real numbers is that they
are ordered.

Order Properties

On the real number line, the positive numbers are to the **right** of 0 and
the negative numbers are to the **left** of 0. See Figure 1.1.

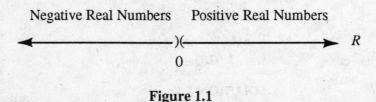

Negative Real Numbers   Positive Real Numbers

0

Figure 1.1

Definitions:
1. Given two unequal real numbers a and b, then a **is less than** b,
   denoted by $a < b$, if and only if $b - a$ is positive.
2. Given two unequal real numbers c and d, then c **is greater than**
   d, denoted by $c > d$, if and only if $c - d$ is positive.

From the above definitions, note that if $a < b$, then a is to the left of b
on the real number line. If $c > d$, then c is to the right of d on the real
number line.

Definition:
The two real numbers p and q are **equal**, denoted by $p = q$, if they
are graphed at the same point on the real number line.

Trichotomy Property

Given any two real numbers a and b, exactly one of the following statements is true:

- a is less than b, denoted by $a < b$.
- a is greater than b, denoted by $a > b$.
- a is equal to b, denoted by $a = b$.

EXERCISE 1.3

Order each of the following pairs of real numbers.

a) $-6, -4$

b) $3, 1$

c) $-8, -6$

d) $-3, 0$

e) $0, 2$

f) $0.12, 0.039$

g) $0.5, \dfrac{1}{4}$

h) $\sqrt{13}, 2.78$

i) $-2.9, -\pi$

j) $\sqrt{4}, 2.0$

SOLUTION 1.3

a) $-6 < -4$

b) $3 > 1$

c) $-8 < -6$

d) $-3 < 0$

e) $0 < 2$

f) $0.12 > 0.039$

g) $0.5 > \dfrac{1}{4}$

h) $\sqrt{13} > 2.78$

i) $-2.9 > -\pi$

j) $\sqrt{4} = 2.0$

Properties of Equality

In addition to the Trichotomy Property, we have the following properties for equality:

Reflexive Property: Let a be a real number. Then $a = a$.

Symmetric Property: Let c and d be real numbers. If $c = d$, then $d = c$.

Transitive Property: Let p, q, and r be real numbers. If $p = q$ and $q = r$, then $p = r$.

Substitution Property: Let u and v be real numbers. If $u = v$, then either may be replaced by the other in any statement or exercise without changing the statement or exercise.

EXERCISE 1.4

Identify the property of equality illustrated by each of the following statements.

a)   If $a = b + c$, then $b + c = a$.

b)   If $pq = rs$ and $rs = tu$, then $pq = tu$.

c)   If $r = -3$ and $r + s = 7$, then $s = 10$.

d) $u + 4 = u + 4$

SOLUTION 1.4

a)   Symmetric

b)   Transitive

c)   Substitution

d)   Reflexive

1.3 FIELD PROPERTIES OF THE REAL NUMBERS

The properties of the real numbers introduced in this section are known as the **field properties**. Emphasis will be placed on the operations of addition and multiplication. The set R, together with the operations of addition and multiplication and the relation of equality, satisfies the field properties.

Closure Properties

For addition. The **sum** of two real numbers is a unique real number. That is, if a and b are real numbers, then $a + b$ is a unique real number. We say that the set R is **closed** under the operation of addition.

For subtraction. The **difference** of two real numbers is a unique real number. That is, if c and d are real numbers, then $c - d$ and $d - c$ are unique real numbers. We say that the set R is **closed** under the operation of subtraction.

For multiplication. The **product** of two real numbers is a unique real number. That is, if c and d are real numbers, then cd is a unique real number. We say that the set R is **closed** under the operation of multiplication.

The set R is not closed under the operation of division since we cannot divide a real number by 0 and get a unique real number. However, the set of **nonzero** real numbers is **closed** under the operation of division.

EXERCISE 1.5

Determine if the set $\{-1, 0, 1\}$ is closed under the operation of:

a) addition

b) subtraction

c) multiplication

d) division

SOLUTION 1.5

a) We have:

$$-1 + 0 = -1, \text{ which is in the set.}$$
$$-1 + 1 = 0, \text{ which is in the set.}$$
$$0 + 1 = 1, \text{ which is in the set.}$$

But

$$-1 + (-1) = -2, \text{ which is not in the set.}$$

Therefore, the given set is **not** closed under the operation of addition.

b) We have:

$$(-1) - (-1) = 0, \text{ which is in the set.}$$
$$0 - 1 = -1, \text{ which is in the set.}$$

But

$$(-1) - 1 = -2, \text{ which is not in the set.}$$

Therefore, the given set is **not** closed under the operation of subtraction.

c) We have:

$$(-1)(0) = 0, \text{ which is in the set.}$$
$$(0)(1) = 0, \text{ which is in the set.}$$
$$(-1)(-1) = 1, \text{ which is in the set.}$$
$$(-1)(1) = -1, \text{ which is in the set.}$$
$$(0)(0) = 0, \text{ which is in the set.}$$
$$(1)(1) = 1, \text{ which is in the set.}$$

Therefore, the given set **is** closed under the operation of multiplication.

d) The given set is **not** closed under the operation of division since $(-1) \div 0$ is not a real number. Therefore, the result is not in the given set.

Commutative Properties

For addition. The order in which we add two real numbers is not important. That is, if a and b are real numbers, then $a + b = b + a$.

For multiplication. The order in which we multiply two real numbers is not important. That is, if c and d are real numbers, then $cd = dc$.

Note that neither the operation of subtraction nor division is a commutative operation. The order in which we subtract or divide is important.

EXERCISE 1.6

Using the real numbers -4 and 5, show that the operations of subtraction and division are not commutative operations.

SOLUTION 1.6

If the operation of subtraction is commutative, then $(-4) - 5 = 5 - (-4)$. We have

$$(-4) - 5 = -9$$

and

$$5 - (-4) = 9.$$

Since $-9 \neq 9$, the operation of subtraction is not commutative.

If the operation of division is commutative, then $(-4) \div 5 = 5 \div (-4)$.

We have

$$(-4) \div 5 = -0.8$$

and

$$5 \div (-4) = -1.25.$$

Since $-0.8 \neq -1.25$, the operation of division is not commutative.

Associative Properties
For addition. The order in which we group real numbers for addition is not important. That is, if a, b, and c are real numbers, then
$(a+b) + c = a + (b+c)$.
For multiplication. The order in which we group real numbers for multiplication is not important. That is, if p, q, and r are real numbers, then $(pq)r = p(qr)$.

Note that neither the operation of subtraction nor division is an associative operation. The order in which we group real numbers for subtraction or division is important.

EXERCISE 1.7

Using the real numbers -3, 2, and 5, show that the operations of subtraction and division are not associative operations.

SOLUTION 1.7

If the operation of subtraction is associative, then $(-3-2) - 5 = -3 - (2-5)$.
We have
$$(-3-2) - 5 = -5 - 5 = -10$$
and
$$-3 - (2-5) = -3 - (-3) = 0.$$
Since $-10 \neq 0$, the operation of subtraction is **not** an associative operation.

If the operation of division is associative, then $(-3 \div 2) \div 5 = -3 \div (2 \div 5)$.
We have
$$(-3 \div 2) \div 5 = -1.5 \div 5 = -0.3$$
and
$$-3 \div (2 \div 5) = -3 \div 0.4 = -7.5.$$
Since $-0.3 \neq -7.5$, the operation of division is **not** an associative operation.

Distributive Properties
Left. If a, b, and c are real numbers, then $a(b+c) = ab + ac$.
Right. If p, q, and r are real numbers, then $(p+q)r = pr + qr$.

EXERCISE 1.8

Using the distributive properties, verify that each of the following is true:

a) $-2(-4+7) = (-2)(-4) + (-2)(7)$

b) $(-5+9)(-3) = (-5)(-3) + (9)(-3)$

SOLUTION 1.8

a) $-2(-4+7) = (-2)(3)$

$$= -6$$

$$-2(-4) + (-2)(7) = 8 + (-14)$$

$$= -6$$

Therefore, $-2(-4+7) = (-2)(-4) + (-2)(7)$.

b) $(-5+9)(-3) = 4(-3)$

$$= -12$$

$$(-5)(-3) + (9)(-3) = 15 + (-27)$$

$$= -12$$

Therefore, $(-5+9)(-3) = (-5)(-3) + (9)(-3)$.

Identity Properties

For Addition: There exists a unique real number 0 such that $a+0 = 0+a$, where a is any real number.

For Multiplication: There exists a unique real number 1 such that $(b)(1) = (1)(b)$, where b is any real number.

Inverse Properties

For Addition: For each real number a, there exists a unique real number $-a$ such that $a + (-a) = (-a) + a = 0$. The real number $-a$ is called the **additive** inverse of the real number a.

For Multiplication: For each *nonzero* real number b, there exists a unique real number $\dfrac{1}{b}$ such that $b\left(\dfrac{1}{b}\right) = \left(\dfrac{1}{b}\right)b = 1$. The real number $\dfrac{1}{b}$ is called the **multiplicative inverse** of the *nonzero* real number b.

EXERCISE 1.9

For each of the following sets, list the additive identity, if it exists, or indicate that none exists.

a) W

b) J

c) The set of all nonpositive integers.

d) The set of all rational numbers less than 3.

SOLUTION 1.9

a) 0

b) None, since 0 is not irrational.

c) 0

d) 0

EXERCISE 1.10

For each of the following sets, list the multiplicative identity, if it exists, or indicate that none exists.

a) N

b) J

c) The set of all rational numbers less than -2.

d) $\{-2, -1, 0, 1, 2\}$

SOLUTION 1.10

a) 1

b) None

c) None

d) 1

EXERCISE 1.11

For each of the following statements, indicate which field property of real numbers is being used.

a) $4 + (-3)$ is a real number

b) $-7 + 0 = -7$

c) $3 + 2 = 2 + 3$

d) $5 + (-5) = 0$

e) $(2 \times 3) \times 4 = 2 \times (3 \times 4)$

f) $\left(\dfrac{-1}{3}\right)(-3) = 1$

g) $(-6)(-5 + 11) = (-6)(-5) + (-6)(-11)$

SOLUTION 1.11

a)   Closure property for addition.

b)   Additive identity property.

c)   Commutative property for addition.

d)   Additive inverse property.

e)   Associative property for multiplication.

f)   Multiplicative inverse property.

g)   Distributive property.

1.4 INTERVALS AND INEQUALITIES

An algebraic equation is the equality of two algebraic expressions. If the equality symbol (=) in an equation is replaced by any of the inequality symbols (>, <, ≥, or ≤), we then have what is called an inequality. Just as is the case with equations, to solve an inequality means to determine all values that can be substituted for the variable(s) to make a true statement. The set of all such values is called the solution set. Often, the solution set is given as an interval.

Intervals

> **Definition**:
> Let *a* and *b* be any two real numbers such that a < b. An **interval** from *a* to *b* is the set of all real numbers between the two real num- bers *a* and *b* and one, neither, or both of the values of *a* and *b*, called the **endpoints** of the interval.

From the above definition, note that an interval can contain neither of its endpoints, exactly one of its endpoints, or both of its endpoints.

> **Definitions:**
> 1. An **open interval** is an interval that contains neither of its end-points. If $a < b$, then the open interval from a to b is denoted by the symbol (a, b). The endpoints are not included.
> 2. A **closed interval** is an interval that contains both of its endpoints. If $a < b$, then the closed interval from a to b is denoted by the symbol $[a, b]$. Both endpoints are included.
> 3. If $a < b$, then the **half-open, half-closed interval** from a to b is denoted by the symbol $(a, b]$. It contains the right endpoint, b, but not the left endpoint, a.
> 4. If $a < b$, then the **half-closed, half-open** interval from a to b is denoted by the symbol $[a, b)$. It contains the left endpoint, a, but not the right endpoint, b.

The various intervals from a to b are illustrated in Figure 1.2.

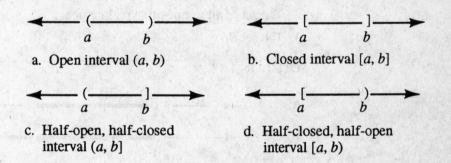

a. Open interval (a, b) b. Closed interval $[a, b]$

c. Half-open, half-closed interval $(a, b]$ d. Half-closed, half-open interval $[a, b)$

Figure 1.2

EXERCISE 1.12

Write each of the following as intervals.
a) The set of all real numbers greater than or equal to -2 and also less than 7.
b) The set of all real numbers greater than 0 but less than 11.
c) The set of all real numbers that are greater than or equal to -10 but less than or equal to 15.
d) The set of all real numbers that are greater than $\sqrt{2}$ but less than or equal to $\sqrt{23}$.

SOLUTION 1.12

a) $[-2, 7)$

b) (0, 11)

c) [–10, 15]

d) $(\sqrt{2}, \sqrt{23}]$

Sometimes, we wish to write sets of real numbers that extend beyond all real bounds. For instance, to write the set of all real numbers that are greater than 5, we use an open interval with the infinity symbol (∞) and write $(5, \infty)$. The infinity symbol is used for convenience; it is not a real number. Similarly, the interval $(-\infty, 9]$ represents the set of all real numbers that are less than or equal to 9.

EXERCISE 1.13

Graph each of the following.

a) (–3, 7]

b) [0, ∞)

c) [–7, 11]

d) (–∞, 0]

e) The set of all positive real numbers.

f) The set of all nonpositive real numbers.

g) The set R.

SOLUTION 1.13

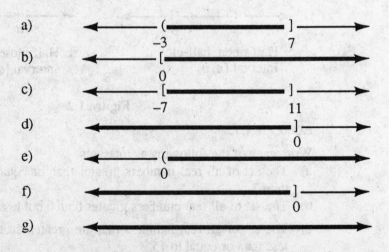

Inequalities

Earlier, we noted that if a and b are real numbers, then either $a < b$, $a = b$, or $a > b$. Each of the statements "$a < b$" and "$a > b$" is called an **inequality**. The symbols "$<$" and "$>$" are called the inequality signs and are always used together with two distinct (unequal) real number symbols. We also use either of these inequality symbols together with the symbol for equality.

The statement $a \leq b$ means that $a < b$ or $a = b$ and is read "a is less than or equal to b." Similarly, the statement $a \geq b$ means that $a > b$ or $a = b$ and is read "a is greater than or equal to b." In a statement of inequality, the inequality sign always points towards the smaller of the two real numbers.

Definition:

Two inequalities are said to be of the **same** or **like** sense if the inequality signs point in the same direction. Two inequalities are said to be of the **opposite** or **unlike** sense if the inequality signs point in different directions.

EXERCISE 1.14

Determine whether the following pairs of inequalities are of the same (like) or opposite (unlike) sense.

a) $x > 2$ and $x < 3$

b) $y > 5$ and $y \geq 8$

c) $4 < 9$ and $-3 < 0$

d) $2y - 3 > 0$ and $3 - 4y < 0$

e) $u^2 \geq 0$ and $u \leq 8$

f) $x^3 - 1 \leq 0$ and $x < x + 1$

SOLUTION 1.14

a)   Opposite

b)   Same

c)   Same

d)   Opposite

e)   Opposite

f)   Same

> **Definition:**
> An inequality that involves only constants or that is true for all permissible values of the variables involved is called an **absolute** or **unconditional inequality**. An inequality that is true for some but not all of the permissible values of the variables involved is called a **conditional inequality**.

The following are all examples of absolute inequalities:

- $4 < 7$
- $-3 > -5$
- $x^2 + 1 > 0$
- $y < y + 1$
- $-p^2 \leq 0$

The following are all examples of conditional inequalities:

- $b + 2 > 3$ (which is true only if $b > 1$)
- $x^2 \leq 4$ (which is true only if $-2 \leq x \leq 2$)
- $2p + 3 \leq p - 1$ (which is true only if $p \leq -4$)
- $\dfrac{u - 1}{2} \leq 6$ (which is true only if $u \leq 13$)

The solution of a conditional inequality involves determining all of those values of the variables for which the inequality is true. We will now discuss various solutions for conditional inequalities. In this section, our attention will be focused on those inequalities involving a single variable.

To solve an inequality, there are certain operations that can be performed which are similar to those performed on equations. These operations depend on properties that will now be introduced. These properties will be stated using the symbol "<." However, the same properties are also true using the other inequality symbols (>, ≤, and ≥).

> **Properties of Inequality**
>
> For the following properties, a and b represent real numbers.
>
> **Positive Multiplication Property**: If $a < b$ and c is a *positive* real number, then $ac < bc$.
>
> **Negative Multiplication Property**: If $a < b$ and c is a *negative* real number, then $ac > bc$.
>
> **Addition Property**: If $a < b$ and c is *any* real number, then $a + c < b + c$.
>
> **Transitivity Property**: If $a < b$ and $b < c$, then $a < c$, where c is *any* real number.

It should be noted that similar properties exist for the operations of division and subtraction. Thus, we can restate the properties of inequalities as follows:

- If both sides of an inequality are multiplied or divided by the same *positive* real number, the resulting inequality has the *same* sense as the original inequality.

- If both sides of an inequality are multiplied or divided by the same *negative* real number, the resulting inequality has the *opposite* sense as the original inequality.

- If the *same* real number is added to or subtracted from both sides of an inequality, the resulting inequality has the *same* sense as the original inequality.

Solving Inequalities   Using the above properties of inequality, we now determine the solution of inequalities in a single variable. The solution sets will be given using interval notation.

EXERCISE 1.15

Solve the inequality $3x - 4 < 5$.

SOLUTION 1.15

$3x - 4 < 5$

$3x < 9$ (Add 4 to both sides; same sense.)

$x < 3$ (Divide both sides by 3, which is positive; same sense.)

The solution is $(-\infty, 3)$.

EXERCISE 1.16

Solve the inequality $6x - 7 > 2x + 9$.

SOLUTION 1.16

$6x - 7 > 2x + 9$

$4x - 7 > 9$ (Subtract $2x$ from both sides; same sense.)

$4x > 16$ (Add 7 to both sides; same sense.)

$x > 4$ (Divide both sides by 4, which is positive; same sense.)

The solution is $(4, +\infty)$.

EXERCISE 1.17

Solve the inequality $2 - 3y \leq 9$.

SOLUTION 1.17

$2 - 3y \leq 9$

$-3y \leq 7$ (Subtract 2 from both sides; same sense.)

$y \geq \dfrac{-7}{3}$ (Divide both sides by -3, which is negative; opposite sense.)

The solution is $\left[\dfrac{-7}{3}, +\infty \right)$.

EXERCISE 1.18

Solve the inequality $-2(u - 3) > 3(1 - u)$.

SOLUTION 1.18

$-2(u - 3) > 3(1 - u)$

$-2u + 6 > 3 - 3u$ (Distributive Property)

$u + 6 > 3$ (Add $3u$ to both sides; same sense.)

$u > -3$ (Subtract 6 from both sides; same sense.)

The solution is $(-3, +\infty)$.

EXERCISE 1.19

Solve the inequality $\dfrac{-6(p + 4)}{5} \geq \dfrac{3 - 2p}{4}$.

SOLUTION 1.19

$\dfrac{-6(p + 4)}{5} \geq \dfrac{3 - 2p}{4}$

$-24(p + 4) \geq 5(3 - 2p)$ (Multiply both sides by 20, which is positive; same sense.)

$$-24p - 96 \geq 15 - 10p$$ (Distributive Property)

$$-14p - 96 \geq 15$$ (Add $10p$ to both sides; same sense.)

$$-14p \geq 111$$ (Add 96 to both sides; same sense.)

$$p \leq \frac{-111}{14}$$ (Divide both sides by -14, which is negative; opposite sense.)

The solution is $\left(-\infty, \dfrac{-111}{14}\right]$.

Consider the two inequalities $-2 \leq 2t + 3$ and $2t + 3 < 7$. These inequalities are of the same sense and can be combined as
$$-2 \leq 2t + 3 < 7.$$
Such an inequality can be solved by working with both inequalities simultaneously, as illustrated in the following exercises.

EXERCISE 1.20

Solve the inequality $-2 \leq 2t + 3 < 7$.

SOLUTION 1.20

$$-2 \leq 2t + 3 < 7$$

$$-5 \leq 2t < 4$$ (Subtract 3 from all three members; same sense.)

$$\frac{-5}{2} \leq t < 2$$ (Divide all three members by 2, which is positive; same sense.)

The solution is $\left[\dfrac{-5}{2}, 2\right)$.

EXERCISE 1.21

Solve the inequality $-4 \leq 3 - 5p \leq 12$.

SOLUTION 1.21

$$-4 \leq 3 - 5p \leq 12$$

$$-7 \leq -5p \leq 9$$ (Subtract 3 from all three members; same sense.)

$$\frac{7}{5} \geq p \geq \frac{-9}{5}$$ (Divide all three members by -5, which is negative; opposite sense.)

The solution is $\left[\dfrac{-9}{5}, \dfrac{7}{5}\right]$.

EXERCISE 1.22

Solve the inequality $3 < \dfrac{2 - 6u}{2} \leq 8$.

SOLUTION 1.22

$$3 < \frac{2 - 6u}{2} \leq 8$$

$6 < 2 - 6u \leq 16$ (Multiply all three members by 2, which is positive; same sense.)

$4 < -6u \leq 14$ (Subtract 2 from all three members; same sense.)

$\dfrac{-2}{3} > u \geq \dfrac{-7}{3}$ (Divide all three members by -6, which is negative; opposite sense.)

The solution is $\left[\dfrac{-7}{3}, \dfrac{-2}{3} \right)$.

1.5 POLYNOMIAL AND OTHER INEQUALITIES

> **Definition:**
> A **polynomial inequality** is an inequality that contains only polynomials.

The following are examples of polynomial inequalities:
- $2x < 3$
- $y^2 + 2 \geq 3y$
- $(2u - 1)(4 - u) \leq 0$

> **Definition:**
> A polynomial inequality is said to be in **standard form** if it is written with 0 on one side of the inequality.

The following are examples of polynomial inequalities in standard form.
- $2x - 3 < 0$
- $y^2 - 3y + 2 \geq 0$
- $(2u - 1)(4 - u) \leq 0$

It should be noted that a polynomial inequality can always be (re)written in standard form by adding an appropriate expression to both sides of the inequality, or subtracting an appropriate expression from both sides.

We now illustrate a method for solving polynomial inequalities.

To solve the inequality $2x^2 - 3x \leq -1$, we follow the procedure given below.

Step 1:

Rewrite the inequality in standard form:

$$2x^2 - 3x + 1 \leq 0$$

Step 2:

Write the equation *associated* with the inequality by replacing the inequality symbol with the equality symbol:

$$2x^2 - 3x + 1 = 0$$

Step 3.

Solve the equation given in Step 2. In this case, we use the method of factoring.

$$(2x - 1)(x - 1) = 0$$

Thus, the solutions are $\frac{1}{2}$ and 1. These become what are called critical values or boundary points.

Definition:

A **critical value** or **boundary point** of an algebraic expression that is a product of two or more factors is any value of the variable(s) that will cause a factor of the product to be equal to 0.

Step 4:

The critical values or boundary points, $\frac{1}{2}$ and 1, divide the set of all real numbers into disjoint subsets, which in this case will be $\left(-\infty, \frac{1}{2}\right)$, $\left(\frac{1}{2}, 1\right)$, and $(1, +\infty)$. It can be determined that on each of these intervals the algebraic sign of the polynomial expression will be the same for all values of x in the interval.

Step 5:

We now determine the algebraic sign of each of the factors, $2x - 1$ and $x - 1$:

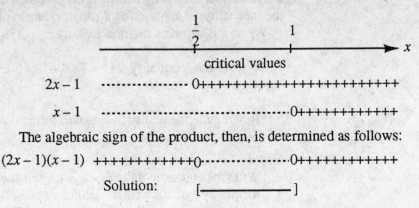

The algebraic sign of the product, then, is determined as follows:

$(2x - 1)(x - 1)$ +++++++++++0···············0+++++++++++

Solution: [————————]

Step 6:

Determine the solution of the inequality by looking at the sign of the polynomial for all real values of x as determined above.

The solution for $2x^2 - 3x + 1 \leq 0$ is $\left[\dfrac{1}{2}, 1\right]$.

For the remainder of our discussion, we will use the expression critical value instead of boundary point. However, you may use either term.

EXERCISE 1.23

Solve the inequality $y(y - 3)(y + 2) \leq 0$.

SOLUTION 1.23

The inequality is already in standard form. The associated equation is
$$y(y - 3)(y + 2) = 0.$$
Since the left-hand side of the equation is in factored form, we determine the solutions of the equation, and hence the critical values, to be 0, 3, and –2. Proceeding, we have:

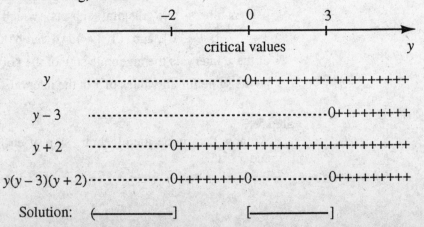

Solution: (————————]   [————————]

The solution is $(-\infty, -2] \cup [0, 3]$.

In writing the solution for the inequality in Exercise 1.23 the symbol "$\cup$" was used. It denotes the union of the two intervals. Hence, the solution $[-\infty, -2) \cup [0, 3]$ means that the solution is the set of all real numbers that are in the interval $(-\infty, -2]$ or in the interval $[0, 3]$.

EXERCISE 1.24

Solve the inequality $w^2 > 5w - 6$.

SOLUTION 1.24

Step 1:

    Rewrite the inequality in standard form.
$$w^2 - 5w + 6 > 0$$

Step 2:

    Determine the associated equation.
$$w^2 - 5w + 6 = 0$$

Step 3:

    Solve the equation.
$$w^2 - 5w + 6 = 0$$
$$(w - 2)(w - 3) = 0$$

The solutions are 2 and 3. Hence, the critical values are 2 and 3.

Step 4:

    Using the critical values, determine the solution for the given inequality.

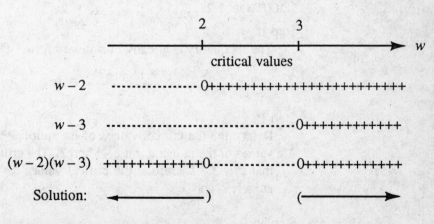

The solution is $(-\infty, 2) \cup (3, +\infty)$.

Rational Inequalities

> **Definition:**
> A **rational inequality** is an inequality that contains rational expressions.

The following are examples of rational inequalities:

- $\dfrac{x+2}{x} > 0$

- $\dfrac{3}{y} \le \dfrac{y}{3}$

- $\dfrac{(u+2)(u-5)}{u+1} < 0$

> **Definition:**
> A **critical value** of an algebraic expression that is a quotient is any value of the variable(s) that will cause the numerator or the denominator of the quotient to be 0.

To solve a rational inequality, we use the same basic procedure as is used to solve polynomial inequalities. However, the critical values are determined differently.

EXERCISE 1.25

Solve the inequality $\dfrac{(t+1)(t-2)}{t-3} < 0$.

SOLUTION 1.25

Step 1:

The inequality is already in standard form. Determine the associated equation.
$$\frac{(t+1)(t-2)}{t-3} = 0$$

Step 2:

Determine the critical values of the rational expression. The critical values of the numerator are −1 and 2. The critical value of the denominator is 3. Therefore, the critical values for the rational expression are −1, 2, and 3.

Step 3:

Using the critical values, solve the inequality.

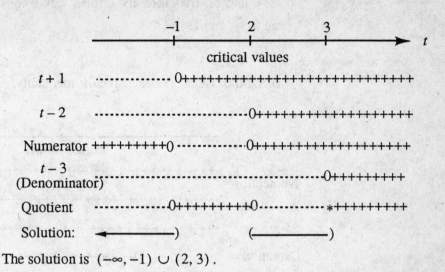

The solution is $(-\infty, -1) \cup (2, 3)$.

Note that in this exercise the symbol "*" was used to denote the value of the quotient at $t = 3$. At $t = 3$, the quotient is not defined since we have division by 0. The symbol "*" will be used when an expression is not defined at a particular value of the given variable.

EXERCISE 1.26

Solve the inequality $\dfrac{2}{y} < \dfrac{3}{2-y}$.

SOLUTION 1.26

Step 1:

Rewrite the inequality in standard form.

$$\frac{2}{y} - \frac{3}{2-y} < 0$$

Step 2:

Combine the expressions on the left-hand side.

$$\frac{2(2-y) - 3y}{y(2-y)} < 0$$

$$\frac{4 - 2y - 3y}{y(2-y)} < 0$$

$$\frac{4 - 5y}{y(2-y)} < 0$$

Step 3:

Determine the critical values for the rational expression. The critical value for the numerator is $\dfrac{4}{5}$. The critical values for the denominator

are 0 and 2. Therefore, the critical values for the rational expression
are 0, $\frac{4}{5}$, and 2.

Step 4:

Using the critical values, solve the inequality.

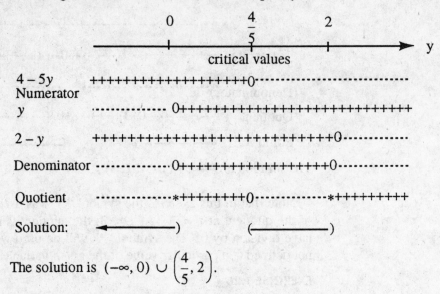

The solution is $(-\infty, 0) \cup \left(\frac{4}{5}, 2\right)$.

1.6 ABSOLUTE VALUES

We now introduce the idea of absolute value of real numbers.

Definition:
Let a be a real number. Then the **absolute value** of a, denoted by $|a|$, is defined by $|a| = a$, if $a \geq 0$, and $|a| = -a$, if $a < 0$.

Hence, according to the definition given above, $|5| = 5$, $|0| = 0$, and $|-7| = 7$.

The absolute value of a real number is also called its **numerical value**, or the value of the real number without regard to its algebraic sign.

EXERCISE 1.27

Determine the value of each of the following.

a) $|3.4|$

b) $|-2.7|$

c) $|7 - 11|$

d) $|\pi - 3|$

e) $|2 - \pi|$

f) $|2y|$, if $y > 0$

g) $|3u|$, if $u < 0$

h) $|2x - 1|$, if $x \geq \dfrac{1}{2}$

i) $|3y - 2|$, if $y < \dfrac{2}{3}$

j) $|u^2 + 3|$

SOLUTION 1.27

a) $|3.4| = 3.4$ since $3.4 > 0$.

b) $|-2.7| = -(-2.7) = 2.7$ since $-2.7 < 0$.

c) $|7 - 11| = |-4| = 4$.

d) $|\pi - 3| = \pi - 3$ since $\pi - 3 > 0$.

e) $|2 - \pi| = -(2 - \pi) = \pi - 2$ since $2 - \pi < 0$.

f) If $y > 0$, then $2y > 0$ and $|2y| = 2y$.

g) If $u < 0$, then $3u < 0$ and $|3u| = -3u$.

h) If $x \geq \dfrac{1}{2}$, then $2x - 1 \geq 0$ and $|2x - 1| = 2x - 1$.

i) If $y < \dfrac{2}{3}$, then $3y - 2 \leq 0$ and $|3y - 2| = -(3y - 2) = 2 - 3y$.

j) $|u^2 + 3| = u^2 + 3$ since $u^2 + 3 > 0$ for all real values of u.

The absolute value of a real number can also be thought of as the undirected distance between the real number and 0 on the real number line. Figure 1.3 illustrates this point.

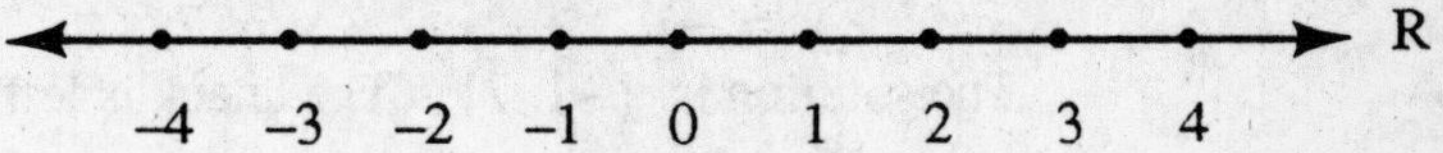

$$\xrightarrow{\hspace{2cm}} R$$
$$-4 \quad -3 \quad -2 \quad -1 \quad 0 \quad 1 \quad 2 \quad 3 \quad 4$$

Figure 1.3

- 3 is three units to the right of 0; hence $|3| = 3$

- -2 is two units to the left of 0; hence $|-2| = 2$

- 0 is zero units from 0; hence $|0| = 0$

- $\dfrac{1}{2}$ is one-half unit to the right of 0; hence $\left|\dfrac{1}{2}\right| = \dfrac{1}{2}$

- -2.5 is two and one-half units to the left of 0; hence $|-2.5| = 2.5$

Equations Involving Absolute Values

Using the definition for the absolute value of a real number, we now solve equations containing absolute value symbols. The solutions of these equations will be given using set notation. If there are no real values satisfying the equation, we will denote that by using the symbol "$\{\ \ \}$" or the symbol "$\emptyset$" to represent the empty or null set. The **empty**, or **null**, **set** is the set containing no elements.

EXERCISE 1.28

Solve the equation $|x| = 4$.

SOLUTION 1.28

If x is a real number and its absolute value is 4, then x must be equal to 4 or the opposite of 4. Therefore, the solution is $\{-4, 4\}$.

EXERCISE 1.29

Solve the equation $|x - 3| = 2$.

SOLUTION 1.29

If $x - 3$ represents a real number and its absolute value is 2, then $x - 3$ must be equal to 2 or the opposite of 2. Hence,

$$x - 3 = 2 \quad \text{or} \quad x - 3 = -2$$
$$x = 5 \qquad\qquad x = 1$$

The solution is $\{1, 5\}$.
Check:

If $x = 1$, then $|x - 3| = |1 - 3| = |-2| = 2$.
If $x = 5$, then $|x - 3| = |5 - 3| = |2| = 2$.

EXERCISE 1.30

Solve the equation $|3 - y| = 4$.

SOLUTION 1.30

$$3 - y = 4 \quad \text{or} \quad 3 - y = -4$$
$$-y = 1 \qquad\qquad -y = -7$$
$$y = -1 \qquad\qquad y = 7$$

The solution is $\{-1, 7\}$. The check is left to the student.

EXERCISE 1.31

Solve the equation $|3u - 5| = -7$.

SOLUTION 1.31

If u is a real number, then $3u - 5$ is a real number. But the absolute value of a real number cannot be negative. Hence, there are no real values of u that satisfy the given equation. The solution is $\emptyset$.

EXERCISE 1.32

Solve the equation $\left|\dfrac{m-2}{3}\right| = 4$.

SOLUTION 1.32

$$\dfrac{m-2}{3} = 4 \quad \text{or} \quad \dfrac{m-2}{3} = -4$$

$$m - 2 = 12 \qquad\qquad m - 2 = -12$$

$$m = 14 \qquad\qquad m = -10$$

The solution is $\{-10, 14\}$.

Sometimes, equations involve absolute values on both sides. We have the following property that may be used.

Property

If a and b represent real numbers and $|a| = |b|$, then $a = b$ or $a = -b$.

EXERCISE 1.33

Solve the equation $|x + 1| = |3 - 2x|$.

SOLUTION 1.33

Using the above property, we have:

$$x + 1 = 3 - 2x \quad \text{or} \quad x + 1 = -(3 - 2x)$$

$$3x + 1 = 3 \qquad\qquad x + 1 = -3 + 2x$$

$$3x = 2 \qquad\qquad 1 = -3 + x$$

$$x = \dfrac{2}{3} \qquad\qquad 4 = x$$

The solution is $\left\{\dfrac{2}{3}, 4\right\}$.

EXERCISE 1.34

Solve the equation $|y - 1| = |1 - y|$.

SOLUTION 1.34

$$y - 1 = 1 - y \quad \text{or} \quad y - 1 = -(1 - y)$$

$$2y - 1 = 1 \qquad\qquad y - 1 = -1 + y$$

$$2y = 2 \qquad\qquad -1 = -1$$

$$y = 1$$

From the equation $y - 1 = 1 - y$, we get the solution $\{1\}$. However, from the equation $y - 1 = -(1 - y)$, we get $-1 = -1$, which is an identity. Hence, the equation is true for all real values of y. Further, note that $y - 1$ is the opposite of $1 - y$ and that their absolute values are equal. The solution, then, is $(-\infty, +\infty)$.

EXERCISE 1.35

Solve the equation $|x + 2| = |x + 4|$.

SOLUTION 1.35

$$x + 2 = x + 4 \quad \text{or} \quad x + 2 = -(x + 4)$$
$$2 = 4 \qquad\qquad\qquad x + 2 = -x - 4$$
$$\text{Not true.} \qquad\qquad 2x + 2 = -4$$
$$2x = -6$$
$$x = -3$$

The solution is $\{-3\}$.

EXERCISE 1.36

Solve the equation $\left|\dfrac{2p - 3}{5}\right| = \left|\dfrac{1 - 4p}{2}\right|$.

SOLUTION 1.36

$$\frac{2p - 3}{5} = \frac{1 - 4p}{2} \qquad \text{or} \qquad \frac{2p - 3}{5} = -\left(\frac{1 - 4p}{2}\right)$$

$$2(2p - 3) = 5(1 - 4p) \qquad\qquad 2(2p - 3) = -5(1 - 4p)$$
$$4p - 6 = 5 - 20p \qquad\qquad\qquad 4p - 6 = -5 + 20p$$
$$24p - 6 = 5 \qquad\qquad\qquad\qquad -6 = -5 + 16p$$
$$24p = 11 \qquad\qquad\qquad\qquad\quad -1 = 16p$$
$$p = \frac{11}{24} \qquad\qquad\qquad\qquad\quad \frac{-1}{16} = p$$

The solution is $\left\{\dfrac{-1}{16}, \dfrac{11}{24}\right\}$.

Inequalities Involving Absolute Values

The equation $|x| = 5$ means that x is 5 units away from 0 on the real number line. Hence, $x = 5$ or $x = -5$ and the solution is $\{-5, 5\}$.

In a similar manner, the inequality $|x| < 5$ means that x is less than 5 units away from 0 on the real number line. That is, $-5 < x < 5$ (see Figure 1.4). The inequality $|x| > 5$ means that x is more than 5 units away from 0 on the real number line. That is, $x > 5$ or $x < -5$ (see Figure 1.5).

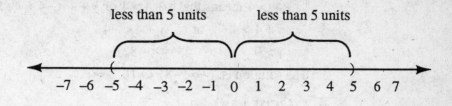

Figure 1.4

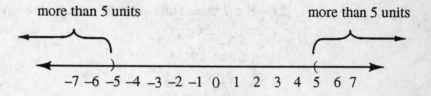

Figure 1.5

Properties of Absolute Value Inequalities

We now summarize some basic properties involving absolute value inequalities. Let x be a real number and let $a > 0$.

- $|x| < a$ means that $-a < x < a$.
- $|x| \leq a$ means that $-a \leq x \leq a$.
- $|x| > a$ means that $x > a$ or $x < -a$.
- $|x| \geq a$ means that $x \geq a$ or $x \leq -a$.

EXERCISE 1.37

Solve the inequality $|x - 3| < 2$.

SOLUTION 1.37

$|x - 3| < 2$ means that $-2 < x - 3 < 2$. We have:

$$-2 < x - 3 < 2$$
$$1 < x < 5.$$

The solution is $(1, 5)$.

EXERCISE 1.38

Solve the inequality $|y + 4| > 4$.

SOLUTION 1.38

$|y + 4| > 4$ means that $y + 4 > 4$ or $y + 4 < -4$. We have:

$$y + 4 > 4 \quad \text{or} \quad y + 4 < -4$$

$$y > 0 \qquad\qquad y < -8$$

The solution is $(-\infty, -8) \cup (0, +\infty)$.

EXERCISE 1.39

Solve the inequality $|2x - 3| \leq 7$.

SOLUTION 1.39

$|2x - 3| \leq 7$ means that $-7 \leq 2x - 3 \leq 7$. We have:

$$-7 \leq 2x - 3 \leq 7$$

$$-4 \leq 2x \leq 10$$

$$-2 \leq x \leq 5.$$

The solution is $[-2, 5]$.

EXERCISE 1.40

Solve the inequality $|4x - 3| \geq 9$.

SOLUTION 1.40

$|4x - 3| \geq 9$ means that $4x - 3 \geq 9$ or $4x - 3 \leq -9$. We have:

$$4x - 3 \geq 9 \quad \text{or} \quad 4x - 3 \leq -9$$

$$4x \geq 12 \qquad\qquad 4x \leq -6$$

$$x \geq 3 \qquad\qquad x \leq \frac{-3}{2}$$

The solution is $\left(-\infty, \dfrac{-3}{2}\right] \cup [3, +\infty)$.

EXERCISE 1.41

Solve the inequality $|3u + 2| < -4$.

SOLUTION 1.41

Since $3u + 2$ represents a real number and the absolute value of a real number cannot be negative, there are no real values of u that will satisfy this inequality. Hence, the solution is $\emptyset$.

In this chapter, we reviewed the real numbers and their basic properties. Interval notation and solutions of inequalities were introduced. Absolute values were reviewed. Both equations and inequalities involving absolute values were examined.

SUPPLEMENTAL EXERCISES

For Exercises 1–4, let N, W, Z, Q, J, and R denote the sets of natural numbers, whole numbers, integers, rational numbers, irrational numbers, and real numbers, respectively.

1. Which of the above sets are closed under the operation of subtraction?
2. Which of the above sets are closed under the operation of division?
3. Which of the above sets have an additive identity element in the set? Identify the additive identity, if it exists.
4. Which of the above sets have a multiplicative identity element in the set? Identify the multiplicative identity, if it exists.
5. If a is a real number, then $-a$ is called the negative of a. What is another name for $-a$?
6. If b is a real number and $b \neq 0$, then $\frac{1}{b}$ is called the multiplicative inverse of b. What is another name for $\frac{1}{b}$?

For Exercises 7–10, state the field property of real numbers being illustrated.

7. $(-2)(-4 + 5) = (-2)(-4) + (-2)(5)$
8. $(-3)(4 \times 7) = (-3)(7 \times 4)$
9. $(7) + (-7) = (-7) + (7) = 0$
10. -6 and -8 are integers, and $(-6) + (-8)$ is an integer.

For Exercises 11-13, write each of the sets as intervals.

11. The set of all real numbers that are less than or equal to -8.
12. The set of all real numbers.
13. The set of all real numbers that are greater than or equal to $\sqrt{3}$ but less than $\sqrt{31}$.

For Exercises 14–16, solve the given inequalities.

14. $4 - 5y \leq 9$
15. $3t + 2 > 4$
16. $-1 \leq 2p - 7 < 6$

For Exercises 17–20, solve the given equations and inequalities.

17. $|3p - 7| = 5$
18. $|1 - 2u| = |3u + 4|$
19. $|4w - 1| \leq 3$
20. $|5 - 3p| > 1$

ANSWERS TO SUPPLEMENTAL EXERCISES

1. Z, Q, and R
2. Q
3. W, Z, Q, and R; 0
4. N, W, Z, Q, and R; 1
5. Opposite of a, or additive inverse of a.
6. Reciprocal of b.
7. Distributive property.
8. Commutative property for multiplication.
9. Additive inverse property.
10. Closure property for addition for the set of integers.
11. $(-\infty, -8]$
12. $(-\infty, +\infty)$
13. $[\sqrt{3}, \sqrt{31})$
14. $[-1, +\infty)$
15. $\left(\dfrac{2}{3}, +\infty\right)$
16. $\left[3, \dfrac{13}{2}\right)$
17. $\{\dfrac{2}{3}, 4\}$
18. $\{-5, \dfrac{-3}{5}\}$
19. $\left[\dfrac{-1}{2}, 1\right]$
20. $\left(-\infty, \dfrac{4}{3}\right) \cup (2, +\infty)$

2

Functions

*T*he unifying theme of this text is the concept of a function. In every-day situations, one hears expressions such as "the function of the air conditioner is to provide comfort," "the function of a speedometer is to indicate how fast one is traveling," "interest is a function of principal, rate, and time," and "the area of a plane circular region is a function of its radius." The mathematical use of the word function has evolved to denote the relationship existing between two variables. In this chapter, we introduce the concept of a function, both as a mapping and as a special set of ordered pairs.

2.1 MAPPINGS

The mathematician who is primarily interested in the area of mathematics known as algebra defines a function as a mapping (sometimes called a map). A **mapping** consists of three items:

- a set called the **domain**,
- a set called the **image set** (or **codomain**), and
- a **rule of correspondence** that associates each element in the domain with a *unique* element in the image set.

The rule of correspondence may be given in the form of an equation, it may be represented in tabular form, or it may be given in words.

Functions as Mappings

> **Definition**
> A **function** f from the set X into the set Y is a mapping that assigns to each element x in X a unique element y in Y. The set X is called the **domain** of the function. The element y is called the **image** of x under the function f. The set of all images of elements in X is called the **range** of the function.

Before examining various examples of mappings, we will emphasize the following implications of the above definition:

- Every element of X has to be paired with a corresponding element in Y.
- If x is an element in X, then there is exactly one element y in Y that is paired with x.
- Not every element in the image set Y has to be used under the mapping f. That is, there may exist some elements in Y that are not images for elements in X.
- It is possible that two or more elements in X will be paired with the same element in Y.

Function Notation

The symbol D_f will denote the domain of the function f.

The symbol R_f will denote the range of the function f.

The symbol $f(x)$ represents the unique element in Y that corresponds to the element x under the function f. Hence, we have $y = f(x)$.

Relative to the symbol $f(x)$, apart from the parentheses, you should see three distinct symbols, namely,

x, which denotes the **domain element**, or the **independent variable value**, or the input element;

$f(x)$, which denotes the **range element**, or the **dependent variable value**, or the output element;

f, which denotes the function itself.

EXERCISE 2.1

Let $X = \{1, 2, 3, 4\}$ and $Y = \{a, b, c\}$. Define f by the following rule of association:

$$
\begin{array}{c}
f \\
\underline{X \quad Y} \\
1 \to a \\
2 \to b \\
3 \to b \\
4 \to c
\end{array}
$$

Is f a function (mapping) from X into Y?

SOLUTION 2.1

We can conclude that f is a function (mapping) from X into Y for the following reasons:

- $D_f = \{1, 2, 3, 4\} = X$; that is, all of the elements in X have been used.
- $R_f = \{a, b, c\} = Y$; that is, all of the images of elements in X under the function f are in Y.
- Each element x in X is paired with a unique (i.e., one and only one) element in Y.

Observe that the element b in Y is associated, or paired, with both elements 2 and 3 in D_f. This does *not* contradict the uniqueness part of the definition of a mapping.

The mapping f from X into Y can also be denoted by the diagram in Figure 2.1.

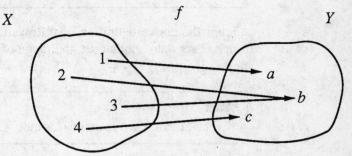

Figure 2.1
The Mapping f from X into Y.

EXERCISE 2.2

Let $A = \{-2, -1, 0, 1, 2\}$ and $B = \{-2, -1, 0, 1, 2, 3, 4, 5\}$.
Define g by the equation $g(a) = a^2$ for all a in A.
Is g a function (mapping) from A into B?

SOLUTION 2.2

We have
-2 is in A and $g(-2) = (-2)^2 = 4$, which is in B.
-1 is in A and $g(-1) = (-1)^2 = 1$, which is in B.
0 is in A and $g(0) = (0)^2 = 0$, which is in B.
1 is in A and $g(1) = (1)^2 = 1$, which is in B.
2 is in A and $g(2) = (2)^2 = 4$, which is in B.

$D_g = \{-2, -1, 0, 1, 2\} = A$. Hence, all of A has been used for the domain of the function g.

$R_g = \{0, 1, 4\}$. Hence, all of the images of elements in A, under the function g, are contained in B.

Each element in A is associated, or paired, with exactly one image element in B.

Therefore, g is a function from A into B.

Onto Mappings

In Exercise 2.1, the range of the function f is the entire image set Y. However, in Exercise 2.2, we observe the range of the function g is not the entire set B.

Definition

A mapping f from a set X into a set Y is said to be a mapping from X **onto** Y, if every element of the set Y is the image of some element of X.

From the above definition, it follows that if a function is a mapping from one set onto another set, the range of the function and the image set are equal.

Fact

Every function is a mapping (or map) from its domain onto its range.

EXERCISE 2.3

Consider the diagram below.

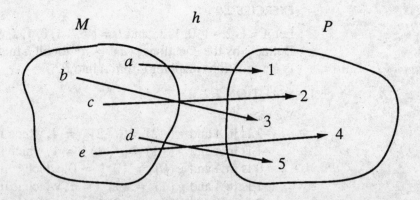

Is h a mapping from M onto P?

SOLUTION 2.3

- $D_h = \{a, b, c, d, e\} = M$. Hence, all of M is used for the domain of h.
- $R_h = \{1, 2, 3, 4, 5\} = P$. Hence, all of the image set P is used for the range of h.
- Each element in M is paired with exactly one element in P.

Hence, h is a mapping from M into P. Since the range and the image set are equal, h is a mapping from M onto P.

EXERCISE 2.4

Consider the mapping F from P into Q defined according to the diagram below.

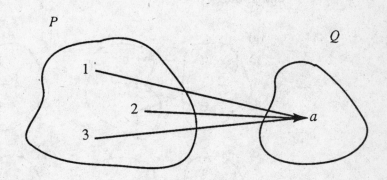

Compute:
a) $F(1)$
b) $F(2)$
c) $F(3)$

SOLUTION 2.4

a) $F(1) = a$
b) $F(2) = a$
c) $F(3) = a$

One-to-One Mappings

In Exercise 2.4, we see that all of the domain elements are associated, or paired, with the same image element. However, F is a mapping from P into Q. Further, F is a mapping from P onto Q. In Exercise 2.3, we note that the images are all different. The mapping in Exercise 2.3 is of a special kind.

Definition:

If *f* is a mapping from *X* into *Y* such that each image element in *Y* is used only once, then *f* is a **one-to-one** mapping.

Note:

A mapping that is one-to-one does *not* have to be onto its image set.

EXERCISE 2.5

Determine which of the following diagrams represent mappings that are one-to-one.

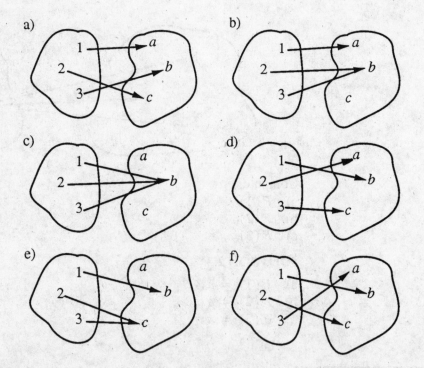

a)   b)   c)   d)   e)   f)

SOLUTION 2.5

(a), (d), and (f)

EXERCISE 2.6

The following table displays the batting averages for five professional baseball players. Does this represent a one-to-one map from the set of players into the set of averages?

Players	Averages
1	.302
2	.297
3	.286
4	.267
5	.297

SOLUTION 2.6

No, since the batting average .297 (an image element) is the same for both Players 2 and 5. That is, one image element is associated with, or paired with, two distinct domain elements. The function, however, is from the set of players onto the set of averages.

2.2 SETS OF ORDERED PAIRS

A function may also be defined as a special set of ordered pairs. This is the approach used by the mathematician whose primary interest lies in the area of mathematics known as analysis. First, we introduce some preliminaries.

Ordered Pairs

Definition:
An **ordered pair** is a pairing of *two* objects such as x and y, written within parentheses and separated by a comma. It is represented by the symbols (x, y) or (y, x). The first object written within the parentheses is called the **first component** of the ordered pair; the second object is called the **second component**.

The components of an ordered pair may be different or they may be alike. For instance, (a, a), (a, b), $(3, 5)$, $(4, 4)$, $(1, a)$, and $(p, 7)$ are all examples of ordered pairs.

Equality of Ordered Pairs

Definition:
Two ordered pairs, (a, b) and (c, d), are **equal**, denoted by $(a, b) = (c, d)$, if and only if $a = c$ and $b = d$.

Cartesian Products

Given two sets, we can construct a new set whose elements are ordered pairs. All of the components of the ordered pairs are elements of the given sets.

> **Definition:**
>
> The **Cartesian product** of two sets, A and B, denoted by $A \times B$, is the set of all ordered pairs of the form (a, b) such that the first component, a, is an element of A and the second component, b, is an element of B.

Observe that the Cartesian product of two given sets is formed by pairing each element of one set, in turn, with every element of the other set. Hence, in the above definition, if set A has m elements and set B has n elements, then the set $A \times B$ has mn elements.

EXERCISE 2.7

If $C = \{1, 2, 3\}$ and $D = \{a, b\}$, form:
a) $C \times D$
b) $D \times C$
c) $C \times C$
d) $D \times D$

SOLUTION 2.7

a) $C \times D = \{ (1, a), (1, b), (2, a), (2, b), (3, a), (3, b) \}$
b) $D \times C = \{ (a, 1), (a, 2), (a, 3), (b, 1), (b, 2), (b, 3) \}$
c) $C \times C = \{(1, 1), (1, 2), (1, 3), (2, 1), (2, 2), (2, 3), (3, 1), (3, 2), (3, 3)\}$
d) $D \times D = \{ (a, a), (a, b), (b, a), (b, b) \}$

The graph of $C \times D$ is given in Figure 2.2 as a set of six points. The set C is graphed on the horizontal axis and the set D is graphed on the vertical axis. Each element in C is paired, in turn, with every element in the set D. The six ordered pairs thus formed are then graphed as indicated.

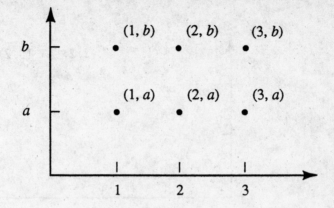

Figure 2.2
Graph of $C \times D$ (A Set of Six Points)

The graph of $D \times C$ is given in Figure 2.3.

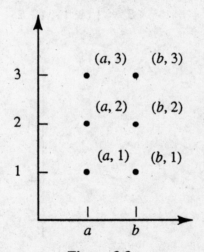

Figure 2.3
Graph of $D \times C$ (A Set of Six Points)

EXERCISE 2.8

Let $E = \{1, 2, 3, 4\}$ and $F = \{2\}$. Form and graph:

a) $E \times F$

b) $F \times E$

SOLUTION 2.8

a) $E \times F = \{ (1,2), (2,2), (3,2), (4,2) \}$

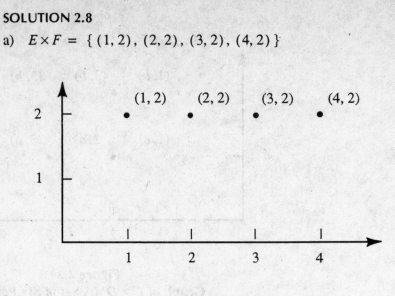

Figure 2.4
Graph of $E \times F$ (A Set of Four Points)

b) $F \times E = \{ (2,1), (2,2), (2,3), (2,4) \}$

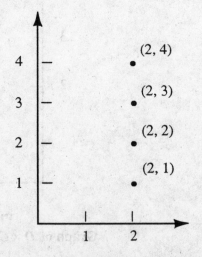

Figure 2.5
Graph of $F \times E$ (A Set of Four Points)

2.3 FUNCTIONS AS SETS OF ORDERED PAIRS

A set of ordered pairs may contain two ordered pairs with the same first components but different second components. In this section, we examine those sets of ordered pairs that have distinct first components.

Definition:

A **function** is a set of ordered pairs such that each of the first components of the set is paired with exactly one second component. The set of all the distinct first components is called the **domain** of the function. The set of all the distinct second components is called the **range** of the function.

EXERCISE 2.9

Determine which of the following sets of ordered pairs are functions. For those that are functions, give their domain and range.

a) $F = \{(1, 2), (2, 3), (2, 5), (4, 4)\}$

b) $G = \{(1, 1), (2, 2), (3, 3)\}$

c) $H = \{(-2, 1), (-3, 1), (0, 1)\}$

d) $J = \{(\sqrt{4}, 2), (\sqrt{4}, -2)\}$

SOLUTION 2.9

a) F is not a function. The first component, 2, is paired with the second components 3 and 5.

b) G is a function. $D_G = \{1, 2, 3\}$ and $R_G = \{1, 2, 3\}$. Note that each element in the domain is paired with exactly one element in the range. In this case, $D_G = R_G$. This is not a requirement for a function.

c) H is a function. $D_H = \{-2, -3, 0\}$ and $R_H = \{1\}$. Note that all elements in the domain of H are paired with the same element in the range of H. This does not violate our definition for a function.

d) J is not a function. Note that the first component, $\sqrt{4}$, is paired with the second components 2 and –2.

Frequently, functions are given as sets of ordered pairs, using what is known as set-builder notation. For instance, we have $\{(x, y)|y = 2x - 1\}$, which is read as "the set of *all* ordered pairs (x, y) such that $y = 2x - 1$." That is, for all of the ordered pairs, the second component, y, is 1 less than twice the first component, x. The set would contain, among others, the ordered pairs $(0, -1)$, $(2, 3)$, and $(-5, -11)$.

EXERCISE 2.10

Consider $f = \{(x, y) | y = x + 3\}$. If f is a function, determine its domain.

SOLUTION 2.10

Since there is exactly one y value associated with each x value, f is a function. $D_f = (-\infty, +\infty)$ since no matter what real number is assigned to x, the y value will be a real number that is 3 more than that number.

EXERCISE 2.11

Consider $g = \{(u, v) | v = 4\}$. If g is a function, determine its domain.

SOLUTION 2.11

Since there is exactly one v value associated with each u value, g is a function. $D_g = (-\infty, +\infty)$ since no matter what real number is assigned to u, the v value is always 4. Notice that $R_g = \{4\}$.

EXERCISE 2.12

Consider $h = \{(p, q) | q^2 = 4p\}$. If h is a function, determine its domain.

SOLUTION 2.12

We can determine that h is not a function. For instance, if $p = 1$, then $q^2 = 4$ and $q = \pm 2$. Hence, the first component, 1, is paired with two distinct second components.

Graphs of Functions

If a function is given as a set of ordered pairs of real numbers, then its graph can be determined in a coordinate plane. Each ordered pair of real numbers becomes the rectangular coordinates of a point which would be the graph of that ordered pair. The graph of the function, then, would be the set of all the points that are graphs of the ordered pairs contained in the function.

Since each element in the domain of a function is paired with exactly one element in its range, any vertical line in the plane that passes through a point on the graph of the function will pass through no other point on the graph. This is known as the **vertical line test** for the graph of a function.

EXERCISE 2.13

Using the vertical line test, determine which of the following are graphs of functions.

a)   A horizontal line.

b)   An oblique line.

c)   A circle.

d)   The letter A.

e)   The letter S.

f)   The letter V.

g)   The numeral 3.

h)   The symbol "/."

i)   The symbol "<."

j)   The symbol "Δ."

SOLUTION 2.13

Any vertical line in the plane containing the graph of a function can intersect the graph at most once. If it intersects the graph more than once, the
graph is not that of a function. Hence, the figures given in (a), (b), (f), and
(h) are graphs of functions; the others are not.

EXERCISE 2.14

Determine which of the graphs given below are graphs of functions.

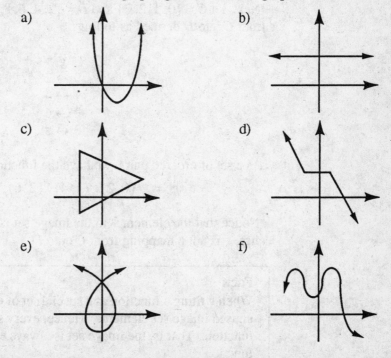

SOLUTION 2.14

Using the vertical line test, we determine that the figures in (a), (b), (d),
and (f) are graphs of functions; the others are not.

A Comparison of the Definitions of a Function

In section 2.1, we defined a function as a mapping from one set into another set. In section 2.3, we defined a function as a special set of ordered pairs. In either case, notice that there is a pairing of objects and that the pairing is ordered. We now compare the two definitions.

Let $A = \{1, 2, 3, 4\}$ and $B = \{a, b, c\}$. Consider the mapping f from A into B defined by the following rule of association:

$$
\begin{array}{c}
f \\
1 \rightarrow a \\
2 \rightarrow b \\
3 \rightarrow b \\
4 \rightarrow c
\end{array}
$$

To rewrite this function as a set of ordered pairs, we note that the function is from A into B. Hence, the first components of the ordered pairs are from A and the second components are from B. We now have

$$f = \{\,(1, a), (2, b), (3, b), (4, c)\,\}$$

as the required set of ordered pairs.

Next, let $C = \{0, 1, 2, 3\}$ and $D = \{2, 4, 6, 8, 9\}$. Consider the mapping g from C into D defined as follows:

$$
\begin{array}{c}
g \\
0 \rightarrow 2 \\
1 \rightarrow 4 \\
2 \rightarrow 6 \\
3 \rightarrow 8
\end{array}
$$

As a set of ordered pairs, we have the function

$$g = \{\,(0, 2), (1, 4), (2, 6), (3, 8)\,\}.$$

Notice that the element 9 in the image set is not used for this mapping. Hence, g is not a mapping from C onto D.

Fact:

When writing a function as a special set of ordered pairs, there are no unused image set elements. Hence, every such function is an onto function. That is, the image set is always equal to the range of the function.

2.4 FUNCTION VALUES AND NATURAL DOMAIN

Function Values

In section 2.1, we made a distinction between the symbol f, which denotes a function, and the symbol $f(x)$, which denotes a function value. The symbol $f(x)$ denotes the second component of the ordered pair of the function with a first component of x. For instance, the symbol $f(3)$ is the second component of the ordered pair of the function with a first component of 3. Hence, the ordered pair $(3, f(3))$ belongs to the function.

EXERCISE 2.15

Let f be the function defined by $f(x) = 3x + 2$. Determine:

a) $f(-3)$

b) $f(0)$

c) $f(2)$

d) $f(3.2)$

e) $f(\sqrt{5})$

SOLUTION 2.15

a) $f(x) = 3x + 2$

     $f(-3) = 3(-3) + 2$ Substituting -3 for x.

           $= -7$

b) $f(x) = 3x + 2$

     $f(0) = 3(0) + 2$ Substituting 0 for x.

        $= 2$

c) $f(x) = 3x + 2$

     $f(2) = 3(2) + 2$ Substituting 2 for x.

        $= 8$

d) $f(x) = 3x + 2$

     $f(3.2) = 3(3.2) + 2$ Substituting 3.2 for x.

          $= 11.6$

e) $f(x) = 3x + 2$

     $f(\sqrt{5}) = 3(\sqrt{5}) + 2$ Substituting $\sqrt{5}$ for x.

        $= 3\sqrt{5} + 2$

EXERCISE 2.16

Let g be the function defined by $g(x) = x^2 - 1$. Determine:

a) $g(-3)$

b) $g(0)$

c) $g(2)$

d) $g(w)$

e) $g(w-1)$

SOLUTION 2.16

a) $g(x) = x^2 - 1$

    $g(-3) = (-3)^2 - 1$ Substituting -3 for x.

          $= 8$

b) $g(x) = x^2 - 1$

    $g(0) = (0)^2 - 1$ Substituting 0 for x.

          $= -1$

c) $g(x) = x^2 - 1$

    $g(2) = (2)^2 - 1$ Substituting 2 for x.

          $= 3$

d) $g(x) = x^2 - 1$

    $g(w) = w^2 - 1$ Substituting w for x, assuming that w is in the domain of g.

e) $g(x) = x^2 - 1$

    $g(w-1) = (w-1)^2 - 1$ Substituting $w-1$ for x. assuming that $w-1$ is in the domain of g.

             $= (w^2 - 2w + 1) - 1$

             $= w^2 - 2w$

EXERCISE 2.17

Use the graph given below to answer the following questions.

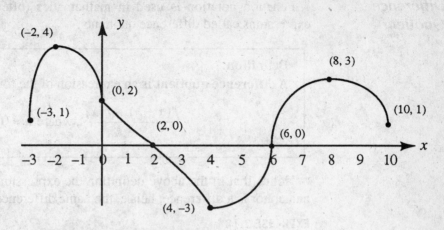

a) Determine $f(-2), f(0),$ and $f(6).$

b) Is $f(10)$ positive or negative?

c) What is the domain of f?

d) What is the range of f?

e) For what values of x is $f(x) \geq 0$?

f) For what values of x is $f(x) < 0$?

g) How many times, if any, does the line with equation $y = 5$ intersect the graph of f?

h) How many times, if any, does the line with equation $y = 2$ intersect the graph of f?

i) How many times, if any, does the line with equation $x = 5$ intersect the graph of f?

j) For what value(s) of x, if any, is $f(x) = 0$?

SOLUTION 2.17

a) $f(-2) = 4; f(0) = 2; f(6) = 0$

b) Positive

c) $D_f = [-3, 10]$

d) $R_f = [-3, 4]$

e) x in $[-3, 2] \cup [6, 10]$

f) x in $(2, 6)$

g) 0

h) 4

i) 1

j) 2 and 6

Difference Quotient

Function notation is used in mathematics (often in the calculus) in expressions called difference quotients.

Definition:

A **difference quotient** is an expression of the form

$$\frac{f(x+h) - f(x)}{h}, \text{ where } h \neq 0.$$

Notice that in the above definition the expression is a quotient and its numerator is a difference; hence, the name difference quotient.

EXERCISE 2.18

Determine the difference quotient for the function f defined by $f(x) = 3x - 5$.

SOLUTION 2.18

Step 1:

We form $f(x + h)$.

$$f(x+h) = 3(x+h) - 5$$
$$= 3x + 3h - 5$$

Step 2:

We subtract $f(x)$ from $f(x+h)$.

$$f(x+h) - f(x) = (3x + 3h - 5) - (3x - 5)$$
$$= 3x + 3h - 5 - 3x + 5$$
$$= 3h$$

Step 3:

We divide the last expression by $h \neq 0$.

$$\frac{f(x+h) - f(x)}{h} = \frac{3h}{h} = 3$$

EXERCISE 2.19

Determine the difference quotient for the function g defined by $g(x) = 4x^2 - 6x + 7$.

SOLUTION 2.19

Step 1:

We form $g(x+h)$.

$$g(x+h) = 4(x+h)^2 - 6(x+h) + 7$$
$$= 4(x^2 + 2hx + h^2) - 6x - 6h + 7$$
$$= 4x^2 + 8hx + 4h^2 - 6x - 6h + 7$$

Step 2:

We subtract $g(x)$ from $g(x+h)$.

$$g(x+h) - g(x) = (4x^2 + 8hx + 4h^2 - 6x - 6h + 7) - (4x^2 - 6x + 7)$$
$$= 4x^2 + 8hx + 4h^2 - 6x - 6h + 7 - 4x^2 + 6x - 7$$
$$= 8hx + 4h^2 - 6h$$

Step 3:

We divide the last expression by $h \neq 0$.

$$\frac{g(x+h) - g(x)}{h} = \frac{8hx + 4h^2 - 6h}{h}$$
$$= 8x + 4h - 6$$

Natural Domain

As we continue to work with functions throughout this text, we will need to determine the domain of a function. Unless otherwise indicated, we use what is called the natural domain.

Definition:

The **natural domain** of a function is the largest subset of the set R that can be used for the domain of the function.

For the remainder of this text, we shall refer to the natural domain of a function simply as its domain.

EXERCISE 2.20

Determine the domain for each of the following functions.

a) $f(x) = 4x + 7$

b) $g(x) = \sqrt{x + 3}$

c) $h(x) = 9$

d) $j(x) = \dfrac{4}{x - 3}$

e) $k(x) = \dfrac{x - 3}{4}$

f) $m(x) = \sqrt{1 - x}$

SOLUTION 2.20

a) $D_f = (-\infty, +\infty)$

b) $g(x)$ is defined only if $x + 3$ is non-negative. Hence, we have

   $x + 3 \geq 0$

   $x \geq -3.$

   Therefore, $D_g = [-3, +\infty).$

c) $D_h = (-\infty, +\infty)$

d) $D_j = (-\infty, 3) \cup (3, +\infty)$ since $j(x)$ is not defined if $x = 3.$

e) $D_k = (-\infty, +\infty)$

f) $D_m = (-\infty, 1]$

EXERCISE 2.21

Determine the domain for each of the following functions.

a) $F(x) = x + 2$

b) $G(x) = \dfrac{x^2 - 4}{x - 2}$

SOLUTION 2.21

a) $D_f = (-\infty, +\infty)$

b) $G(x) = \dfrac{x^2 - 4}{x - 2}$

$\qquad = \dfrac{(x - 2)(x + 2)}{x - 2}$

$\qquad = x + 2 \qquad \text{(if } x \neq 2\text{)}$

$\quad D_G = (-\infty, 2) \cup (2, +\infty)$

In Exercise 2.21, we note that the functions F and G appear to be equal functions. However, that is not true. Their domains are different.

Equal Functions

Definition:

Two functions, f and g, are **equal** if and only if

(a) $D_f = D_g$, and

(b) $f(x) = g(x)$ for *all* x in the common domain.

In this chapter, we introduced the concept of a function, both as a mapping (or map) and as a special set of ordered pairs. One-to-one functions and functions that are onto their image sets were also discussed. A comparison of the two definitions given for a function was also examined.

SUPPLEMENTAL EXERCISES

In Exercises 1–5, for each set of ordered pairs, if a function is defined, a) write its domain, b) write its range, and c) indicate whether the function is one-to-one.

1. $A = \{(-3, 2), (0, 4), (5, 5), (7, -3)\}$
2. $B = \{(-5, 2), (-3, 2), (1, 2), (4, 2)\}$
3. $C = \{(2, -3), (-3, 2), (2, 0), (2, 5), (2, 7)\}$
4. $D = \{(1, 2), (2, 3), (3, 4), (4, 5), (5, 6), (6, 7)\}$
5. $E = \{(-4, 0), (-3, 4), (-3, 1), (0, 2), (1, -4), (5, 0)\}$

In Exercises 6–10, for each set of ordered pairs, if a function is defined, a) write its domain and b) indicate whether the function is one-to-one (x and y represent real numbers).

6. $f = \{(x, y) \mid y = x^2\}$
7. $g = \{(x, y) \mid x = y^2\}$
8. $h = \{(x, y) \mid 2x + 3y = 5\}$
9. $j = \{(x, y) \mid y = 2\}$
10. $k = \{(x, y) \mid x = -3\}$

ANSWERS TO SUPPLEMENTAL EXERCISES

1. A is a function. $D_A = \{-3, 0, 5, 7\}$ and $R_A = \{2, 4, 5, -3\}$. A is one-to-one.

2. B is a function. $D_B = \{-5, -3, 1, 4\}$ and $R_B = \{2\}$. B is not one-to-one.

3. C is not a function.

4. D is a function. $D_D = \{1, 2, 3, 4, 5, 6\}$ and $R_D = \{2, 3, 4, 5, 6, 7\}$. D is one-to-one.

5. E is not a function.

6. f is a function. $D_f = (-\infty, +\infty)$. f is not one-to-one.

7. g is not a function.

8. h is a function. $D_h = (-\infty, +\infty)$. h is one-to-one.

9. j is a function. $D_j = (-\infty, +\infty)$. j is not one-to-one.

10. k is not a function.

3

Special Functions

In this chapter, we examine some special functions that are of interest in mathematics. We also look at some functions that can be formed by combining two or more functions.

3.1 INCREASING, DECREASING, AND CONSTANT-VALUED FUNCTIONS

Consider the function f defined by $y = f(x) = 2x + 5$. The symbols y, $f(x)$, and $2x + 5$ all represent the same quantities. That is, they all represent range values for the function f. As the domain values, x, increase, we note that the range values, $2x + 5$, also increase.

Next, consider the function g defined by $y = g(x) = 1 - 2x$. Again, the symbols y, $g(x)$, and $1 - 2x$ all represent the same range values. As the domain values, x, increase, we note that the range values, $1 - 2x$, decrease.

Finally, consider the function h defined by $y = h(x) = 5$. The symbols y, $h(x)$, and 5 all represent the same range values. As the domain values, x, increase, we note that the range values remain constant, namely, 5.

Definition:

Let f be a function defined on some set S. Then:
- f is said to be **increasing** on S if $f(a) < f(b)$ whenever $a < b$.
- f is said to be **decreasing** on S if $f(a) > f(b)$ whenever $a < b$.
- f is said to be **constant valued** on S if $f(a) = f(b)$ for **every** a and b in S.

Fact:

- If a function is increasing on its domain, then the graph of the function is *rising* as we look at it from left to right over its domain.
- If a function is decreasing on its domain, then the graph of the function is *falling* as we look at it from left to right over its domain.
- If a function is constant valued on its domain, then the graph of the function is a *horizontal line (segment)*.

EXERCISE 3.1

Consider the function F graphed in Figure 3.1. Determine on what interval(s) the function F is:

a)   increasing
b)   decreasing
c)   constant valued.

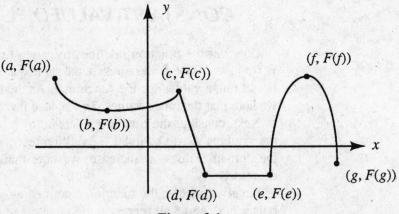

Figure 3.1

SOLUTION 3.1

a)   The function F is increasing on the intervals (b, c) and (e, f).

b)   The function F is decreasing on the intervals (a, b), (c, d), and (f, g).

c)   The function F is constant valued on the interval (d, e).

EXERCISE 3.2

Consider the function G defined by $G(x) = x + 3$ for all x in the interval $[-2, 5]$. Determine whether G is an increasing, decreasing, or constant-valued function on the interval.

SOLUTION 3.2

When $x = -1$, then $G(-1) = (-1) + 3 = 2$. When $x = 2$, then $G(2) = 2 + 3 = 5$. We have that $-1 < 2$ and $G(-1) < G(2)$. You may now be tempted to say that G is an increasing function on the interval. But is it? For G to be an increasing function on the interval, $G(a) < G(b)$ *whenever* a and b are in the interval and $a < b$. Let a and b be two distinct elements in the given interval such that $a < b$. We have:

$a < b$

$a + 3 < b + 3$ Add 3 to both sides; same sense inequalities.

$G(a) < G(b)$ Substitution.

Therefore, G is an increasing function on the interval $[-2, 5]$.

3.2 EVEN AND ODD FUNCTIONS

Functions can be classified as being even, odd, or neither.

Definitions:
- A function f is **even** if $f(-x) = f(x)$ for all x in its domain.
- A function f is **odd** if $f(-x) = -f(x)$ for all x in its domain.

From the above definitions, we can conclude that a function does not have to be either even or odd.

EXERCISE 3.3

Determine whether each of the following functions is even, odd, or neither.

a) $f(x) = 3x$

b) $g(x) = x^2 + 2$

c) $h(x) = 4x - 1$

d) $p(x) = 6$

SOLUTION 3.3

a) The function f is odd since

$f(-x) = 3(-x)$

$\qquad = -3x$

$\qquad = -(3x)$

$\qquad = -f(x)$.

b) The function g is even since

$$g(-x) = (-x)^2 + 2$$
$$= x^2 + 2$$
$$= g(x).$$

c) The function h is neither even nor odd since

$$h(-x) = 4(-x) - 1$$
$$= -4x - 1$$
$$= -(4x + 1)$$

and $-(4x + 1)$ is neither $h(x)$ nor $-h(x)$.

d) The function p is even since

$$p(-x) = 6$$
$$= p(x).$$

Symmetry

Consider the graph of a function f in the xy-plane. If the graph of f is "folded" along the y-axis and the portion of the graph to the left of the y-axis coincides with that portion to the right of the y-axis, then the graph is symmetric with respect to the y-axis.

Definition:

The graph of the function f is said to be **symmetric with respect to the y-axis** if, whenever the point (x, y) is on the graph of f, the point $(-x, y)$ is also on the graph of f; that is, if $f(-x) = f(x)$ for all x in the domain of f.

From the above definition, we note that if the graph of a function is symmetric with respect to the y-axis, then the function is an even function.

Definition:

The graph of the function f is said to be **symmetric with respect to the origin** if, whenever the point (x, y) is on the graph of f, the point $(-x, -y)$ is also on the graph of f; that is, if $f(-x) = -f(x)$ for all x in the domain of f.

From the above definition, we note that if the graph of a function is symmetric with respect to the origin, then the function is an odd function.

EXERCISE 3.4

Determine whether the graph of the function F defined by $y = F(x) = x^4 - 5$ is symmetric with respect to the y-axis.

SOLUTION 3.4

If the graph of F is symmetric with respect to the y-axis, then $F(-x)$ must equal $F(x)$ for all x in the domain of F. We have:

$$F(-x) = (-x)^4 - 5$$
$$= x^4 - 5$$
$$= F(x) \text{ for all } x \text{ in the domain of } F.$$

Therefore, the graph of F is symmetric with respect to the y-axis. We also note that the function F is an even function.

EXERCISE 3.5

Determine whether the graph of the function G defined by $y = G(x) = 3x^2 - 2x$ is symmetric with respect to the y-axis.

SOLUTION 3.5

$$G(-x) = 3(-x)^2 - 2(-x)$$
$$= 3x^2 + 2x$$

$G(-x) \neq G(x)$ for all x in the domain of G. Therefore, the graph of G is not symmetric with respect to the y-axis.

EXERCISE 3.6

Determine whether the graph of the function H defined by $y = H(x) = 5x - 7$ is symmetric with respect to the origin.

SOLUTION 3.6

If the graph of H is symmetric with respect to the origin, then $H(-x)$ must equal $-H(x)$ for all x in the domain of H. We have

$$H(-x) = 5(-x) - 7$$
$$= -5x - 7$$
$$= -(5x + 7)$$

$H(-x) \neq -H(x)$ for all x in the domain of H. Therefore, the graph of H is not symmetric with respect to the origin.

EXERCISE 3.7

Determine whether the graph of the function P defined by $y = P(x) = x^3 - 3x$ is symmetric with respect to the origin.

SOLUTION 3.7

$$P(-x) = (-x)^3 - 3(-x)$$
$$= -x^3 + 3x$$
$$= -(x^3 - 3x)$$
$$= -P(x) \text{ for all } x \text{ in the domain of } P$$

Therefore, the graph of the function P is symmetric with respect to the origin.

EXERCISE 3.8

Examine the graphs of the functions in Figure 3.2 to determine which of the functions are even, odd, or neither.

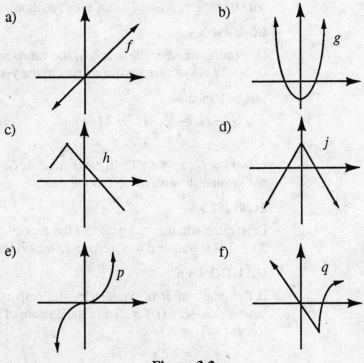

Figure 3.2

SOLUTION 3.8

a) The graph of f appears to be symmetric with respect to the origin. Hence, the function f is odd.

b) The graph of g appears to be symmetric with respect to the y-axis. Hence, the function g is even.

c) The graph of h does not appear to be symmetric with respect to the y-axis or the origin. Hence, the function h is neither even nor odd.

d) The graph of j appears to be symmetric with respect to the y-axis.

Hence, the function *j* is even.

e) The graph of *p* appears to be symmetric with respect to the origin. Hence, the function *p* is odd.

f) The graph of *q* does not appear to be symmetric with respect to the *y*-axis or the origin. Hence, the function *q* is neither even nor odd.

3.3 SOME VERY IMPORTANT FUNCTIONS

In this section, we examine some very important functions. The functions will be defined and then discussed.

Identity Function

> **Definition:**
> The **Identity function**, denoted by *I*, is defined by $y = I(x) = x$ for all *x* in *R*.

EXERCISE 3.9

Consider the Identity function defined above.

a) What is its domain?

b) What is its range?

c) What is its graph?

d) Is the function one-to-one?

e) Is the function increasing, decreasing, or neither?

f) Determine $I(-3)$, $I(0)$, and $I(6)$.

g) Write the function as a set of ordered pairs.

SOLUTION 3.9

a) $D_I = (-\infty, +\infty)$

b) $R_I = (-\infty, +\infty)$

c) The graph of *I* is an oblique line passing through the origin of the plane and bisecting the first and third quadrants. (See Figure 3.3.)

d) The function is one-to-one.

e) The function is increasing on its domain.

f) $I(-3) = -3, I(0) = 0, I(6) = 6$

g) $I = \{ (x, y) \mid y = I(x) = x \text{ for all in } R \}$

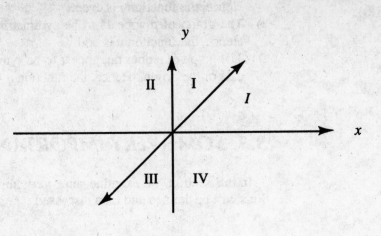

Figure 3.3

The Absolute Value Function

> **Definition:**
> Let x be any real number. Then the **absolute value** of x, denoted by $|x|$, is defined as
> $$|x| = x, \text{ if } x \geq 0$$
> $$|x| = -x, \text{ if } x < 0.$$

From the above definition, we note that the absolute value of every real number is non-negative. That is, the absolute value of a real number cannot be negative.

EXERCISE 3.10

Determine each of the following:

a) $|7.4|$

b) $|-3|$

c) $|2\sqrt{3}|$

d) $|-\pi|$

e) $|y|$, if $y > 0$

f) $|u|$, if $u < 0$

g) $|2.1 - \pi|$

h) $|-2p|$, if $p < 0$

i) $|3r - 2|$, if $r = 0.5$

j) $|r - s|$, if $r < s$

SOLUTION 3.10

a) $|7.4| = 7.4$

b) $|-3| = 3$

c) $|2\sqrt{3}| = 2\sqrt{3}$

d) $|-\pi| = \pi$

e) If $y > 0$, $|y| = y$.

f) If $u < 0$, then $|u| = -u$.

g) $|2.1 - \pi| = \pi - 2.1$

h) If $p < 0$, then $-2p > 0$, and $|-2p| = 2p$.

i) If $r = 0.5$, then $3r - 2 = 3(0.5) - 2 = -0.5$, and
   $|3r - 2| = |-0.5| = 0.5$.

j) If $r < s$, then $r - s < 0$, and $|r - s| = s - r$.

Definition:
> The **Absolute Value function**, denoted by **Abs**, is defined by
> $$y = \text{Abs}(x) = |x| \text{ for all } x \text{ in } R.$$

EXERCISE 3.11

Consider the Absolute Value function defined above.

a) What is its domain?

b) What is its range?

c) What is the graph of the Absolute Value function?

d) Is the function **Abs** one-to-one?

e) For what value(s) of x is the function **Abs** increasing?

f) Determine **Abs** (-6.2); **Abs** (0); and **Abs** (7.03).

g) Rewrite the function **Abs** as a set of ordered pairs.

SOLUTION 3.11

a) $D_{\text{Abs}} = (-\infty, +\infty)$

b) $R_{\text{Abs}} = [0, +\infty)$

c) The graph of **Abs** is the union of two rays, starting at the origin, one of which bisects the first quadrant, and the other, the second quadrant. (See Figure 3.4.)

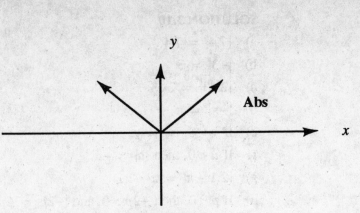

Figure 3.4

d)   **Abs** is not one-to-one. For instance, 4 in the range of **Abs** is paired with both -4 and 4 in the domain of **Abs**.

e)   **Abs** is increasing for x in $(0, +\infty)$.

f)   **Abs** $(-6.2) = 6.2$, **Abs** $(0) = 0$, **Abs** $(7.03) = 7.03$

g)   **Abs** $= \{ (x, y) \mid y = $ **Abs** $(x) = |x|$ for all x in $R \}$

The Greatest Integer Function

> **Definition:**
> Let x represent any real number. Then, by the **greatest integer** in x, denoted by $[x]$, we mean the greatest integer that is less than or equal to x.

EXERCISE 3.12

Determine each of the following:

a) $[3.56]$

b) $[0.98]$

c) $[\sqrt{2}]$

d) $[\pi]$

e) $[-2.6]$

f) $[-0.19]$

g) $[-8.9]$

h) $[3 - 5.2]$

i) $[4 - \pi]$

j) $[-2\pi + 1]$

(Use 3.14 for π.)

SOLUTION 3.12

a) $[3.56] = 3$

b) $[0.98] = 0$

c) $[\sqrt{2}] = 1$

d) $[\pi] = [3.14] = 3$

e) $[-2.6] = -3$

f) $[-0.19] = -1$

g) $[-8.9] = -9$

h) $[3 - 5.2] = [-2.2] = -3$

i) $[4 - \pi] = [4 - 3.14] = [0.86] = 0$

j) $[-2\pi + 1] = [-2(3.14) + 1] = [-6.28 + 1] = [-5.28] = -6$

On the real number line, the greatest integer in a real number is the real number itself, if the number is an integer, or the nearest integer to the left of the number.

Definition:

The **Greatest Integer function**, denoted by G, is defined by
$$y = G(x) = [x] \text{ for all } x \text{ in } R.$$

EXERCISE 3.13

Consider the Greatest Integer function defined above.

a) What is its domain?

b) What is its range?

c) What is the graph of the function?

d) Is the function G one-to-one?

e) For what value(s) of x is the function G decreasing?

f) Determine $G(-5.01)$; $G(0)$; and $G(6.999)$.

g) Rewrite the function G as a set of ordered pairs.

SOLUTION 3.13

a) $D_G = (-\infty, +\infty)$

b) $R_G = Z$ (the set of all integers)

c) The graph of G is the set of horizontal line segments that include their left endpoints but not their right endpoints. (See Figure 3.5.)

d) The function G is not one-to-one. For instance, 4 in the range of G is paired with both 4.1 and 4.87 in the domain of *G*.

e)   There are no values of *x* for which the function *G* is decreasing.

f) $G(-5.01) = -6$; $G(0) = 0$; $G(6.999) = 6$

g) $G = \{ (x, y) \mid y = G(x) = [x] \text{ for all } x \text{ in } R \}$

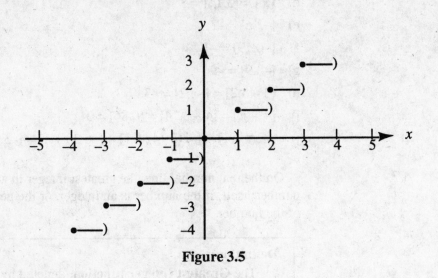

Figure 3.5

Constant-Valued Functions

A function is a Constant-Valued function if and only if all of the second components of its ordered pairs are equal. That is, the range of a constant-valued function contains exactly one element.

> **Definition:**
> A **Constant-Valued function**, denoted by *C*, is defined by
> $y = C(x) = c$ for all *x* in *R*, and such that *c* is a specific real number.

EXERCISE 3.14

Consider the Constant-Valued function defined by $y = C(x) = 7$.

a)   What is its domain?

b)   What is its range?

c)   What is the graph of the function?

d)   Is the function one-to-one?

e)   Determine $C(-6)$; $C(0)$; and $C(3.7)$.

f)   Rewrite the function *C* as a set of ordered pairs.

SOLUTION 3.14

a) $D_C = (-\infty, +\infty)$

b) $R_C = \{7\}$

c) The graph of C is a horizontal line, in the xy-plane, that is 7 units from the x-axis and passes through the point with coordinates $(0, 7)$. (See Figure 3.6.)

d) The function C is not one-to-one. The element 7 in the range of C is paired with all real numbers in the domain of C.

e) $C(-6) = 7; C(0) = 7; C(3.7) = 7$

f) $C = \{ (x, y) | y = C(x) = 7 \text{ for all } x \text{ in } R \}$

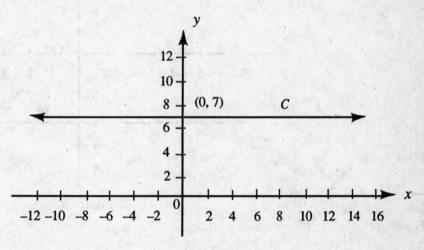

Figure 3.6

EXERCISE 3.15

Consider the Constant-Valued function C defined by $y = C(x) = -3$ for all x in R.

a) What is its domain?

b) What is its range?

c) What is the graph of the function?

d) Is the function one-to-one?

e) Determine $C(-4.9)$; $C(0)$; and $C(9.13)$.

f) Rewrite the function C as a set of ordered pairs.

SOLUTION 3.15

a) $D_C = (-\infty, +\infty)$

b) $R_C = \{-3\}$

c) The graph of C is a horizontal line in the xy-plane that is 3 units below the x-axis and passes through the point with coordinates $(0, -3)$. (See Figure 3.7.)

d) The function C is not one-to-one. The element -3 in the range of C is paired with every element in the domain of C.

e) $C(-4.9) = -3; C(0) = -3; C(9.13) = -3$

f) $C = \{(x, y) | y = C(x) = -3 \text{ for all } x \text{ in } R\}$

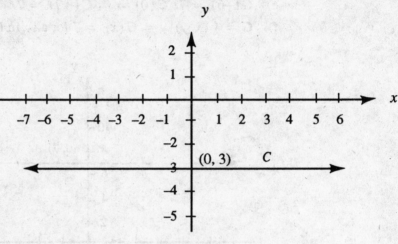

Figure 3.7

Signum Function

Definition:

The **Signum function**, denoted by **Sgn**, is defined by

$$\text{Sgn } (x) = \begin{cases} 1, & \text{if } x > 0 \\ 0, & \text{if } x = 0 \\ -1, & \text{if } x < 0 \end{cases}$$

EXERCISE 3.16

Consider the function **Sgn** defined above.

a) What is its domain?

b) What is its range?

c) What is the graph of the function?

d) Is the function one-to-one?

e) Determine: **Sgn** (–8.7); **Sgn** (0); and **Sgn** (12.75).

f) Rewrite **Sgn** as a set of ordered pairs.

SOLUTION 3.16

a) $D_{Sgn} = (-\infty, +\infty)$

b) $R_{Sgn} = \{-1, 0, 1\}$

c) The graph of **Sgn** consists of two horizontal rays without their end-points—one starting at (0, 1) and extending to the right, and the other starting at (0, –1) and extending to the left—and also the point at the origin. (See Figure 3.8.)

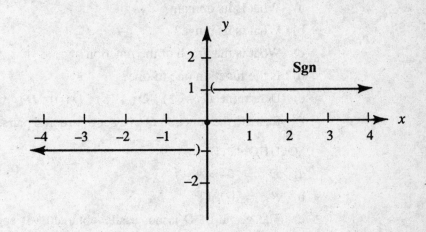

Figure 3.8

d) The function **Sgn** is not one-to-one. The element 1 in its range is paired with both 3 and 7 in its domain.

e) **Sgn** (–8.7) = –1; **Sgn** (0) = 0; and **Sgn** (12.75) = 1

f) **Sgn** = { $(x, y)\,|\,y$ = **Sgn** (x) = –1, if $x < 0$; y = **Sgn** (x) = 0, if $x = 0$; y = **Sgn** (x) = 1, if $x > 0$}

Notice that the Signum function and the function that follows are defined piecewise. That is, the functions are defined differently on different parts of their domains.

The Dirichlet Function

Definition:

The **Dirichlet function**, denoted by D, is defined by

$$D(x) = \begin{cases} 1, & \text{if } x \text{ is rational} \\ 0, & \text{if } x \text{ is irrational} \end{cases}$$

EXERCISE 3.17

Consider the Dirichlet function defined above.

a) What is its domain?

b) What is its range?

c) What is the graph of the function?

d) Is the function one-to-one?

e) Determine $D(-\sqrt{3})$; $D(-1.8)$; $D(0)$; $D(4.023)$; and $D(\sqrt{8})$.

f) Rewrite the function D as a set of ordered pairs.

SOLUTION 3.17

a) $D_D = (-\infty, +\infty)$

b) $R_D = \{0, 1\}$

c) The graph of D is not easily obtained. It appears to look like the graph given in Figure 3.9, consisting of two parallel lines with equations $y = 1$ and $y = 0$. However, only the points on the line $y = 1$ that have rational x values are included. And only the points on the line $y = 0$ that have irrational x values are included.

d) The function D is not one-to-one. For instance, the element 1 in the range of D is paired with both -3 and 0.5 in the domain of D.

e) $D(-\sqrt{3}) = 0$; $D(-1.8) = 1$; $D(0) = 1$; $D(4.023) = 1$; $D(\sqrt{8}) = 0$

f) $D = \{ (x, y) \mid y = D(x) = 1, \text{ if } x \text{ is rational}; y = D(x) = 0, \text{ if } x \text{ is irrational}\}$

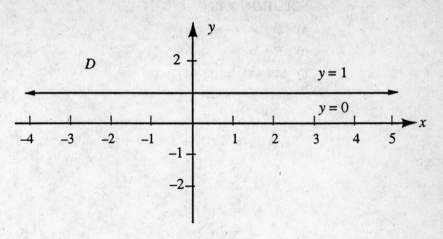

Figure 3.9

3.4 OTHER SPECIAL FUNCTIONS

Linear Functions

A special type of function is a linear function. Linear functions will be introduced briefly in this section and will be discussed in more detail in chapter 5.

> **Definition:**
> A **linear function**, denoted by **L**, is defined by
> $$y = L(x) = mx + b,$$
> where m and b are elements in R and $m \neq 0$.

For particular values of m and b, we will use different symbols to identify linear functions.

EXERCISE 3.18

Consider the function A defined by $y = A(x) = 2x + 1$.

a) What is its domain?

b) What is its range?

c) Determine $A(-4)$; $A(-2)$; $A(0)$; $A(2)$; and $A(4)$.

d) What is the graph of the function?

e) Is the function A one-to-one?

SOLUTION 3.18

a) $D_A = (-\infty, +\infty)$

b) $R_A = (-\infty, +\infty)$

c) $A(-4) = 2(-4) + 1$ $A(-2) = 2(-2) + 1$
         $= -8 + 1$ $= -4 + 1$
         $= -7$ $= -3$

   $A(0) = 2(0) + 1$ $A(2) = 2(2) + 1$
         $= 0 + 1$ $= 4 + 1$
         $= 1$ $= 5$

   $A(4) = 2(4) + 1$
         $= 8 + 1$
         $= 9$

d) The graph of A is the oblique line (i.e., neither horizontal nor vertical) passing through the points with coordinates $(-2, -3)$, $(0, 1)$, and $(2, 5)$. (See Figure 3.10.)

e) From the graph above, it appears that any horizontal line in the xy-plane that intersects the graph of A will intersect it only once. That means that each element in the range of A is paired with exactly one element in the domain of A. Hence, A is one-to-one.

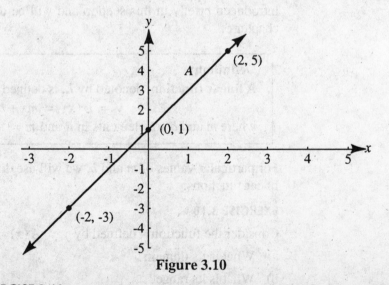

Figure 3.10

EXERCISE 3.19

Consider the function B defined by $y = B(x) = -3x + 1$.

a) What is its domain?

b) What is its range?

c) Determine $B(-2)$; $B(-1)$; $B(0)$; $B(1)$; and $B(2)$.

d) What is the graph of the function?

e) Is the function B one-to-one?

SOLUTION 3.19

a) $D_B = (-\infty, +\infty)$

b) $R_B = (-\infty, +\infty)$

c)
$$
\begin{aligned}
B(-2) &= -3(-2) + 1 & \qquad B(-1) &= -3(-1) + 1 \\
&= 6 + 1 & &= 3 + 1 \\
&= 7 & &= 4 \\
B(0) &= -3(0) + 1 & B(1) &= -3(1) + 1 \\
&= 0 + 1 & &= -3 + 1 \\
&= 1 & &= -2 \\
B(2) &= -3(2) + 1 \\
&= -6 + 1 \\
&= -5
\end{aligned}
$$

d) The graph of B is the oblique line (i.e., neither horizontal nor vertical) passing through the points with coordinates $(-1, 4)$, $(0, 1)$, and $(1, -2)$. (See Figure 3.11.)

e) From the graph above, it appears that any horizontal line in the *xy*-plane that intersects the graph of B will intersect it only once. That means that each element in the range of B is paired with exactly one element in the domain of B. Hence, B is one-to-one.

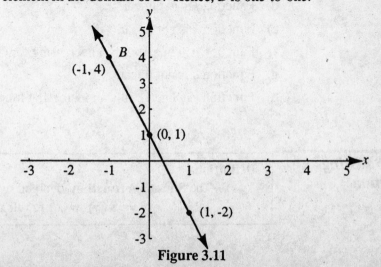

Figure 3.11

The graph of a linear function is an oblique line. The symbol m used in defining a linear function is called the **slope** of the line. If $m > 0$, then the general direction of the line is upward as you move to the right. That is, the linear function is an increasing function. If $m < 0$, then the general direction is downward as you move to the right. Hence, the linear function is a decreasing function.

EXERCISE 3.20

For each of the following, determine whether the given function is linear. If it is, classify it as an increasing or a decreasing function.

a) $f(x) = 7x - 3$

b) $g(x) = 4 - 9x$

c) $h(x) = 5x$

d) $j(x) = 2$

e) $k(x) = \sqrt{x - 1}$

f) $p(x) = \dfrac{2 - x}{3}$

g) $q(x) = \dfrac{3}{x}$

h) $r(x) = x - \dfrac{2}{7}$

SOLUTION 3.20

a) Function f is linear. It is an increasing function since $m = 7 > 0$.

b) Function g is linear. It is a decreasing function since $m = -9 < 0$.

c) Function h is linear. It is an increasing function since $m = 5 > 0$.

d) Function j is not linear.

e) Function k is not linear.

f) Function p is linear. It is a decreasing function since $m = \dfrac{-1}{3} < 0$.

g) Function q is not linear.

h) Function r is linear. It is an increasing function since $m = 1 > 0$.

The Squaring Function

Definition:

The **Squaring function**, denoted by S, is defined by
$$y = S(x) = x^2 \text{ for all } x \text{ in } R.$$

EXERCISE 3.21

Consider the function S defined above.

a) What is its domain?

b) What is its range?

c) Determine $S(-3)$; $S(-2)$; $S(-1)$; $S(0)$; $S(1)$; $S(2)$; and $S(3)$.

d) Determine the graph of the function.

e) Is the function one-to-one?

f) Rewrite the function as a set of ordered pairs.

SOLUTION 3.21

a) $D_S = (-\infty, +\infty)$

b) For all real values of x, x^2 is always non-negative. The minimum value for x^2 is 0 (when $x = 0$). There is no largest value of x^2. Therefore, $R_S = [0, +\infty)$.

c) $S(-3) = (-3)^2 = 9$

$S(-2) = (-2)^2 = 4$

$S(-1) = (-1)^2 = 1$

$S(0) = (0)^2 = 0$

$S(1) = (1)^2 = 1$

$S(2) = (2)^2 = 4$

$S(3) = (3)^2 = 9$

d) From (c) above, we note that the graph of S will pass through the points with coordinates $(-3, 9)$; $(-2, 4)$; $(-1, 1)$; $(0, 0)$; $(1, 1)$; $(2, 4)$; and $(3, 9)$. The graph is called a parabola and is given in Figure 3.12.

e) From the graph given above, we determine that S is not one-to-one. For instance, 4 in the range of S is paired with both -2 and 2 in the domain of S.

f) $S = \{ (x, y) \mid y = S(x) = x^2 \text{ for all } x \text{ in } R \}$

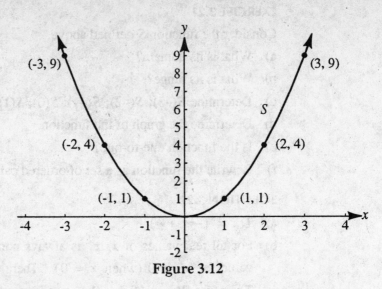

Figure 3.12

EXERCISE 3.22

a) Classify the Squaring function as being even, odd, or neither.

b) Determine whether the graph of the Squaring function is symmetric with respect to the y-axis or the origin.

SOLUTION 3.22

a) $S(-x) = (-x)^2$

$\qquad\quad = x^2$

$\qquad\quad = S(x)$

Therefore, S is an even function.

b) Since S is an even function, its graph is symmetric with respect to the y-axis.

The Square Root Function

> **Definition:**
> The **Square Root function**, denoted by s, is defined by
> $$y = s(x) = \sqrt{x} \text{ for all } x \text{ in } [0, +\infty).$$

EXERCISE 3.23

Consider the function s defined above.

a) What is its domain?

b) What is its range?

c) Determine $s(0)$; $s(1)$; $s(4)$; $s(9)$; $s(16)$; and $s(25)$.

d) Determine the graph of the function.

e) Is the function one-to-one?

f) Rewrite the function s as a set of ordered pairs.

SOLUTION 3.23

a) Since $\sqrt{x}$ is a real number only if $x \geq 0$, then $D_s = [0, +\infty)$.

b) If $x \geq 0$, then $\sqrt{x} \geq 0$. Therefore, $R_s = [0, +\infty)$.

c) $s(0) = \sqrt{0} = 0$

   $s(1) = \sqrt{1} = 1$

   $s(4) = \sqrt{4} = 2$

   $s(9) = \sqrt{9} = 3$

   $s(16) = \sqrt{16} = 4$

   $s(25) = \sqrt{25} = 5$

d) From (c) above, we determine that the graph of s will pass through the points with coordinates $(0, 0)$; $(1, 1)$; $(4, 2)$; $(9, 3)$; $(16, 4)$; and $(25, 5)$. The graph is given in Figure 3.13.

e) From the graph given below, it appears that any horizontal line in the xy-plane that intersects the graph will intersect it only once. That is, each range element is paired with exactly one domain element. Hence, the function s is one-to-one.

f) $s = \{(x, y) \mid y = s(x) = \sqrt{x} \text{ for all } x \text{ in } R\}$

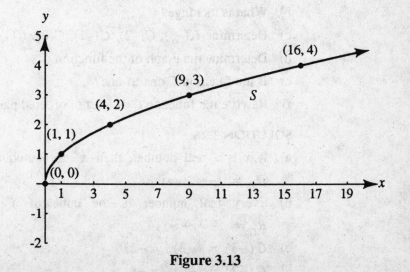

Figure 3.13

EXERCISE 3.24

Determine whether the function s defined above is increasing, decreasing, or neither.

SOLUTION 3.24

As x, in the domain of the function, increases, $\sqrt{x}$ also increases. Therefore, the function s is an increasing function. Also, from the graph of s, we determine that the graph is rising from left to right.

EXERCISE 3.25

Determine whether the graph of s defined above is symmetric with respect to the y-axis, or the origin.

SOLUTION 3.25

$s(-x) = \sqrt{-x}$. If $x > 0$, then $-x < 0$; and $\sqrt{-x}$ is not defined as a real number. Therefore, s is neither odd nor even and the graph of s is not symmetric with respect to the y-axis or the origin.

The Cubing Function

Definition:

The **Cubing function**, denoted by C, is defined by
$$y = C(x) = x^3 \text{ for all } x \text{ in } R.$$

EXERCISE 3.26

Consider the function C defined above.

a) What is its domain?

b) What is its range?

c) Determine $C(-3)$; $C(-2)$; $C(-1)$; $C(0)$; $C(1)$; $C(2)$; and $C(3)$.

d) Determine the graph of the function.

e) Is the function C one-to-one?

f) Rewrite the function C as a set of ordered pairs.

SOLUTION 3.26

a) If x is a real number, then x^3 is a unique real number. Hence,
$$D_C = (-\infty, +\infty).$$

b) Every real number is the cube of a real number. Hence,
$$R_C = (-\infty, +\infty).$$

c) $C(-3) = (-3)^3 = -27$
$C(-2) = (-2)^3 = -8$

$$C(-1) = (-1)^3 = -1$$
$$C(0) = (0)^3 = 0$$
$$C(1) = (1)^3 = 1$$
$$C(2) = (2)^3 = 8$$
$$C(3) = (3)^3 = 27$$

d) From (c), we determine that the graph of C passes through the points with coordinates $(-3, -27)$; $(-2, -8)$; $(-1, -1)$; $(0, 0)$; $(1, 1)$; $(2, 8)$; and $(3, 27)$. The graph is given in Figure 3.14.

e) From the graph of C, it appears that every horizontal line in the xy-plane will intersect the graph of C exactly once. That is, each element in the range of C is paired with exactly one element in the domain. Hence, the function C is one-to-one.

f) $C = \{(x, y) \mid y = C(x) = x^3 \text{ for all } x \text{ in } R\}$

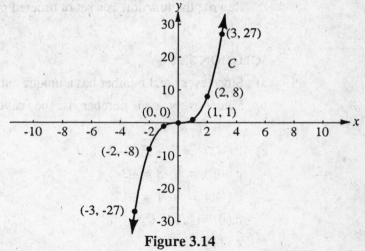

Figure 3.14

EXERCISE 3.27

a) Is the function C defined above even, odd, or neither?

b) Is the graph of the function C symmetric with respect to the y-axis or the origin?

SOLUTION 3.27

a) $C(-x) = (-x)^3$
$$= -x^3$$
$$= -C(x)$$

Therefore, the function C is an odd function.

b) Since the function C is an odd function, its graph is symmetric with respect to the origin.

The Cube Root Function

> **Definition:**
>
> The **Cube Root function**, denoted by c, is defined by
> $$y = c(x) = \sqrt[3]{x} \text{ for all } x \text{ in } R.$$

EXERCISE 3.28

Consider the function c defined above.

a) What is its domain?

b) What is its range?

c) Determine $c(-27)$; $c(-8)$; $c(-1)$; $c(0)$; $c(1)$; $c(8)$; and $c(27)$.

d) Determine the graph of c.

e) Is the function one-to-one?

f) Rewrite the function as a set of ordered pairs.

SOLUTION 3.28

a) Since every real number has a unique cube root, $D_c = (-\infty, +\infty)$.

b) Since every real number is the cube root for a real number, $R_c = (-\infty, +\infty)$.

c) $c(-27) = \sqrt[3]{-27} = -3$

   $c(-8) = \sqrt[3]{-8} = -2$

   $c(-1) = \sqrt[3]{-1} = -1$

   $c(0) = \sqrt[3]{0} = 0$

   $c(1) = \sqrt[3]{1} = 1$

   $c(8) = \sqrt[3]{8} = 2$

   $c(27) = \sqrt[3]{27} = 3$

d) From (c) above, we determine that the graph of c will pass through the points with coordinates $(-27, -3)$; $(-8, -2)$; $(-1, -1)$; $(0, 0)$; $(1, 1)$; $(8, 2)$; and $(27, 3)$. The graph is given in Figure 3.15.

e) From the graph of c, it appears that every horizontal line in the xy-plane will intersect the graph of c exactly once. That is, each element in the range of c is paired with exactly one element in the domain of c. Therefore, the function c is one-to-one.

f) $c = \{ (x, y) \mid y = c(x) = \sqrt[3]{x} \text{ for all } x \text{ in } R \}$

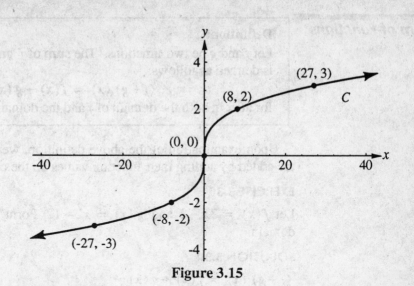

Figure 3.15

EXERCISE 3.29

a) Is the function c defined above even, odd, or neither?
b) Is the graph of the function c symmetric with respect to the y-axis or the origin?

SOLUTION 3.29

a) $c(-x) = \sqrt[3]{-x}$

   $\qquad\quad = -\sqrt[3]{x}$

   $\qquad\quad = -c(x)$

   Therefore, the function c is an odd function.

b) Since the function c is an odd function, then its graph is symmetric with respect to the origin.

3.5 ALGEBRA OF FUNCTIONS

In this section, we consider combining two functions using the operations of addition, subtraction, multiplication, and division.

Sum of Functions

> **Definition:**
>
> Let f and g be two functions. The **sum** of f and g, denoted by $f+g$, is defined as follows:
>
> $$(f+g)(x) = f(x) + g(x)$$
>
> for all x in both the domain of f and the domain of g.

Upon examination of the above definition, we note that two functions are added by adding their function values on the common domains.

EXERCISE 3.30

Let $f(x) = 3x - 2$ and $g(x) = x^2 - 1$. Form $f+g$ and determine its domain.

SOLUTION 3.30

$$
\begin{aligned}
(f+g)(x) &= f(x) + g(x) \\
&= (3x - 2) + (x^2 - 1) \\
&= x^2 + 3x - 3
\end{aligned}
$$

Since $D_f = (-\infty, +\infty)$ and $D_g = (-\infty, +\infty)$, we have $D_{f+g} = (-\infty, +\infty)$.

EXERCISE 3.31

Let $F = 2x^3 + 4$ and $G = \sqrt{2x - 1}$. Form $F + G$ and determine its domain.

SOLUTION 3.31

$$
\begin{aligned}
(F+G)(x) &= F(x) + G(x) \\
&= (2x^3 + 4) + (\sqrt{2x - 1}) \\
&= 2x^3 + 4 + \sqrt{2x - 1}
\end{aligned}
$$

$G(x)$ is meaningful only if $2x - 1 \geq 0$, or $x \geq \dfrac{1}{2}$. Hence, $D_G = \left[\dfrac{1}{2}, +\infty\right)$. Also, $D_F = (-\infty, +\infty)$. Therefore, $D_{F+G} = \left[\dfrac{1}{2}, +\infty\right)$.

EXERCISE 3.32

Let $p(x) = \dfrac{1}{x}$ and $q(x) = \sqrt{x}$. Form $p + q$ and determine its domain.

SOLUTION 3.32

$$
\begin{aligned}
(p+q)(x) &= p(x) + q(x) \\
&= \frac{1}{x} + \sqrt{x}
\end{aligned}
$$

$D_p = (-\infty, 0) \cup (0, +\infty)$ and $D_q = [0, +\infty)$. Therefore, $D_{p+q} =$

$(0, +\infty)$.

EXERCISE 3.33

Let $A(x) = 4x - 7$, $B(x) = \sqrt{x+2}$, $C(x) = \dfrac{1}{x-2}$, and $D(x) = x^2$.
If possible, determine each of the following. If not possible, indicate why not.

a) $(A+B)(2)$

b) $(A+B)(0)$

c) $(A+C)(3)$

d) $(B+C)(2)$

e) $(B+D)(7)$

f) $(C+D)(5)$

g) $(A+B+C)(0)$

h) $(B+C+D)(1)$

SOLUTION 3.33

a) $(A+B)(2) = A(2) + B(2)$
$= [4(2) - 7] + \sqrt{2+2}$
$= 1 + \sqrt{4}$
$= 1 + 2$
$= 3$

b) $(A+B)(0) = A(0) + B(0)$
$= [4(0) - 7] + \sqrt{0+2}$
$= -7 + \sqrt{2}$

c) $(A+C)(3) = A(3) + C(3)$
$= [4(3) - 7] + \dfrac{1}{3-2}$
$= 5 + 1$
$= 6$

d) $(B+C)(2) = B(2) + C(2)$
$= \sqrt{2+2} + \dfrac{1}{2-2}$
$= \sqrt{4} +$ (division by 0; not defined)

Therefore, $(B+C)(2)$ is not defined.

e) $(B+D)(7) = B(7) + D(7)$

$$= \sqrt{7+2} + (7)^2$$

$$= 3 + 49$$

$$= 52$$

f) $(C+D)(5) = C(5) + D(5)$

$$= \frac{1}{5-2} + (5)^2$$

$$= \frac{1}{3} + 25$$

$$= 25\frac{1}{3}$$

g) $(A+B+C)(0) = A(0) + B(0) + C(0)$

$$= [4(0) - 7] + \sqrt{0+2} + \frac{1}{0-2}$$

$$= -7 + \sqrt{2} + \left(\frac{-1}{2}\right)$$

$$= \frac{-15}{2} + \sqrt{2}$$

h) $(B+C+D)(1) = B(1) + C(1) + D(1)$

$$= \sqrt{1+2} + \frac{1}{1-2} + (1)^2$$

$$= \sqrt{3} + (-1) + 1$$

$$= \sqrt{3}$$

Difference of Functions

Definition:

Let f and g be two functions. The **difference** of f and g, denoted by $f-g$, is defined as follows:

$$(f-g)(x) = f(x) - g(x)$$

for all x in both the domain of f and the domain of g.

EXERCISE 3.34

Let $f(x) = 5x^3 - 2$ and $g(x) = 3 - 2x$. Form $f-g$ and determine its domain.

SOLUTION 3.34

$$(f-g)(x) = f(x) - g(x)$$

$$= (5x^3 - 2) - (3 - 2x)$$
$$= 5x^3 - 2 - 3 + 2x$$
$$= 5x^3 + 2x - 5$$

$D_f = (-\infty, +\infty)$ and $D_g = (-\infty, +\infty)$. Therefore, $D_{f-g} = (-\infty, +\infty)$.

EXERCISE 3.35

Let $F(x) = \sqrt{3x - 1}$ and $G(x) = \sqrt{x + 4}$. Form $F - G$ and determine its domain.

SOLUTION 3.35

$$(F - G)(x) = F(x) - G(x)$$
$$= \sqrt{3x - 1} - \sqrt{x + 4}$$

$F(x)$ is meaningful only if $3x - 1 \geq 0$, or $x \geq \dfrac{1}{3}$. $G(x)$ is meaningful

only if $x + 4 \geq 0$, or $x \geq -4$. Therefore, $D_{F+G} = \left[\dfrac{1}{3}, +\infty \right)$.

EXERCISE 3.36

Let $A(x) = -3x$, $B(x) = \dfrac{1}{x}$, $C(x) = \sqrt{3 - 2x}$, and $D(x) = x^2 - 1$.
If possible, determine each of the following. If not possible, indicate why not.

a) $A(-4)$

b) $B(5)$

c) $C(2)$

d) $D(-2)$

e) $(A - C)(-3)$

f) $(B - D)(1)$

g) $(C - B)(0)$

h) $(A - B - D)(4)$

i) $(B - C - D)(-1)$

SOLUTION 3.36

a) $A(-4) = -3(-4)$
$$= 12$$

b) $B(5) = \dfrac{1}{5}$

c) $C(2) = \sqrt{3 - 2(2)}$
$$= \sqrt{3 - 4}$$

$$= \sqrt{-1}$$

which is not a real number. Hence, $C(2)$ is not defined as a real number.

d) $D(-2) = (-2)^2 - 1$

$$= 4 - 1$$

$$= 3$$

e) $(A-C)(-3) = A(-3) - C(-3)$

$$= -3(-3) - \sqrt{3 - 2(-3)}$$

$$= 9 - \sqrt{9}$$

$$= 6$$

f) $(B-D)(1) = B(1) - D(1)$

$$= \frac{1}{1} - [(1)^2 - 1]$$

$$= 1 - 0$$

$$= 1$$

g) $(C-B)(0) = C(0) - B(0)$

$$= \sqrt{3 - 2(0)} - \frac{1}{0}$$

$$= \sqrt{3} - \text{(division by 0; not defined)}$$

Therefore, $(C-B)(0)$ is not defined.

h) $(A-B-D)(4) = A(4) - B(4) - D(4)$

$$= -3(4) - \frac{1}{4} - [(4)^2 - 1]$$

$$= -12 - \frac{1}{4} - 15$$

$$= -27\frac{1}{4}$$

i) $(B-C-D)(-1) = B(-1) - C(-1) - D(-1)$

$$= \frac{1}{-1} - \sqrt{3 - 2(-1)} - [(-1)^2 - 1]$$

$$= -1 - \sqrt{5} - 0$$

$$= -1 - \sqrt{5}$$

Product of Functions

> **Definition:**
>
> Let f and g be two functions. **The product** of f and g, denoted by $f \cdot g$, is defined as follows: $(f \cdot g)(x) = f(x) \cdot g(x)$ for all x in both the domain of f and the domain of g.

EXERCISE 3.37

Let $f(x) = 2 - 6x$ and $g(x) = \dfrac{2}{x-3}$. Form $f \cdot g$ and determine its domain.

SOLUTION 3.37

$(f \cdot g)(x) = f(x) \cdot g(x)$

$$= (2 - 6x)\left(\frac{2}{x-3}\right)$$

$$= \frac{2(2 - 6x)}{x - 3}$$

$$= \frac{4 - 12x}{x - 3}$$

$D_f = (-\infty, +\infty)$. If $x = 3$, then $x - 3 = 0$.
Hence, $D_g = (-\infty, 3) \cup (3, +\infty)$.
Therefore, $D_{f \cdot g} = (-\infty, 3) \cup (3, +\infty)$.

EXERCISE 3.38

Let $F(x) = \sqrt{x + 2}$ and $G(x) = \dfrac{1}{x^2 - 9}$. Form $f \cdot g$ and determine its domain.

SOLUTION 3.38

$(F \cdot G)(x) = F(x) \cdot G(x)$

$$= \sqrt{x + 2}\left(\frac{1}{x^2 - 9}\right)$$

$$= \frac{\sqrt{x + 2}}{x^2 - 9}$$

$F(x)$ is a real number only if $x \geq -2$. Hence, $D_F = [-2, +\infty)$. $G(x)$ is a real number only if $x \neq -3$ or $x \neq 3$. Hence, $D_G = (-\infty, -3) \cup (-3, 3) \cup (3, +\infty)$.
Therefore, $D_{F \cdot G} = [-2, 3) \cup (3, +\infty)$.

EXERCISE 3.39

Let $A(x) = x^2$, $B(x) = \sqrt[3]{3 - x}$, $C(x) = \sqrt{x - 8}$, and $D(x) = \dfrac{-1}{5x}$.

If possible, determine each of the following. If not possible, indicate why not.

a) $(A \cdot B) \, (-2)$

b) $(A \cdot C) \, (4)$

c) $(B \cdot D) \, (0)$

d) $(C \cdot D) \, (9)$

e) $(A \cdot D) \, (-3)$

f) $(B \cdot C) \, (4)$

g) $(A \cdot C \cdot D) \, (12)$

h) $(A \cdot B \cdot D) \, (-5)$

SOLUTION 3.39

a) $(A \cdot B) \, (-2) = A \, (-2) \cdot B \, (-2)$
$$= (-2)^2 \, \sqrt[3]{3 - (-2)}$$
$$= 4 \, \sqrt[3]{5}$$

b) $(A \cdot C) \, (4) = A \, (4) \cdot C \, (4)$
$$= (4)^2 \sqrt{4 - 8}$$
$$= 16 \cdot \text{(not defined, since } 4 - 8 < 0)$$

Therefore, $(A \cdot C) \, (4)$ is not defined.

c) $(B \cdot D) \, (0) = B \, (0) \cdot D \, (0)$
$$= \sqrt[3]{3 - 0} \left(\frac{-1}{(5) \, (0)} \right)$$
$$= \sqrt[3]{3} \cdot \text{(not defined; division by 0)}$$

Therefore, $(B \cdot D) \, (0)$ is not defined.

d) $(C \cdot D) \, (9) = C \, (9) \cdot D \, (9)$
$$= \sqrt{9 - 8} \left(\frac{-1}{(5) \, (9)} \right)$$
$$= (1) \left(\frac{-1}{45} \right)$$
$$= \frac{-1}{45}$$

e) $(A \cdot D) \, (-3) = A \, (-3) \cdot D \, (-3)$
$$= (-3)^2 \left(\frac{-1}{(5) \, (-3)} \right)$$
$$= 9 \left(\frac{1}{15} \right)$$

$$= \frac{3}{5}$$

f) $(B \cdot C)(4) = B(4) \cdot C(4)$

$$= \sqrt[3]{3-4}\sqrt{4-8}$$

$$= \sqrt[3]{-1}\sqrt{-4}$$

$$= (-1) \cdot (\text{not defined})$$

Therefore, $(B \cdot C)(4)$ is not defined.

g) $(A \cdot C \cdot D)(12) = A(12) \cdot C(12) \cdot D(12)$

$$= (12)^2 \sqrt{12-8} \left(\frac{-1}{(5)(12)} \right)$$

$$= (144)(2) \left(\frac{-1}{60} \right)$$

$$= \frac{-24}{5}$$

h) $(A \cdot B \cdot D)(-5) = A(-5) \cdot B(-5) \cdot D(-5)$

$$= (-5)^2 \sqrt[3]{3-(-5)} \left(\frac{-1}{(5)(-5)} \right)$$

$$= (25)(2) \left(\frac{-1}{-25} \right)$$

$$= 2$$

Quotients of Functions

> **Definition:**
>
> Let f and g be two functions. The **quotient** of f and g, denoted by $\frac{f}{g}$, is defined as follows:
>
> $$\left(\frac{f}{g} \right)(x) = \frac{f(x)}{g(x)}$$
>
> for all x in both the domain of f and the domain of g, and such that $g(x) \neq 0$.

EXERCISE 3.40

Let $f(x) = 2x - 1$ and $g(x) = x + 5$. Form $\frac{f}{g}$ and determine its domain.

SOLUTION 3.40

$$\left(\frac{f}{g} \right)(x) = \frac{f(x)}{g(x)}$$

$$= \frac{2x - 1}{x + 5}$$

$D_f = (-\infty, +\infty)$ and $D_g = (-\infty, +\infty)$. However, $g(x) = x + 5 = 0$ when $x = -5$. Therefore, $D_{f/g} = (-\infty, -5) \cup (-5, +\infty)$.

EXERCISE 3.41

For the functions f and g given in Exercise 3.40, form $\frac{g}{f}$ and determine its domain.

SOLUTION 3.41

$$\left(\frac{g}{f}\right)(x) = \frac{g(x)}{f(x)}$$

$$= \frac{x + 5}{2x - 1}$$

Again, $D_f = D_g = (-\infty, +\infty)$. However, $f(x) = 2x - 1 = 0$ when $x = \frac{1}{2}$. Therefore, $D_{g/f} = \left(-\infty, \frac{1}{2}\right) \cup \left(\frac{1}{2}, +\infty\right)$.

EXERCISE 3.42

Let $A(x) = \sqrt{3 - x}$, $B(x) = x^2 - 1$, $C(x) = 7x$, and $D(x) = x^2 + 1$. If possible, determine each of the following. If not possible, indicate why not.

a) $\left(\dfrac{A}{B}\right)(0)$

b) $\left(\dfrac{B}{C}\right)(-1)$

c) $\left(\dfrac{C}{B}\right)(-1)$

d) $\left(\dfrac{B}{D}\right)(2)$

e) $\left(\dfrac{D}{A}\right)(5)$

f) $\left(\dfrac{C}{D}\right)(-2)$

g) $\left(\dfrac{D}{C}\right)(0)$

SOLUTION 3.42

a) $\left(\dfrac{A}{B}\right)(0) = \dfrac{A(0)}{B(0)}$

$$= \frac{\sqrt{3-0}}{(0)^2 - 1}$$

$$= \frac{\sqrt{3}}{-1}$$

$$= -\sqrt{3}$$

b) $\left(\dfrac{B}{C}\right)(-1) = \dfrac{B(-1)}{C(-1)}$

$$= \frac{(-1)^2 - 1}{(7)(-1)}$$

$$= \frac{0}{-7}$$

$$= 0$$

c) $\left(\dfrac{C}{B}\right)(-1) = \dfrac{C(-1)}{B(-1)}$

$$= \frac{(7)(-1)}{(-1)^2 - 1}$$

$$= \frac{-7}{0}$$

which is undefined.

d) $\left(\dfrac{B}{D}\right)(2) = \dfrac{B(2)}{D(2)}$

$$= \frac{(2)^2 - 1}{(2)^2 + 1}$$

$$= \frac{3}{5}$$

e) $\left(\dfrac{D}{A}\right)(5) = \dfrac{D(5)}{A(5)}$

$$= \frac{(5)^2 + 1}{\sqrt{3} - 5}$$

$$= \frac{26}{\text{Not defined}}$$

Hence, $\left(\dfrac{D}{A}\right)(5)$ is not defined.

f) $\left(\dfrac{C}{D}\right)(-2) = \dfrac{C(-2)}{D(-2)}$

$$= \dfrac{(7)(-2)}{(-2)^2 + 1}$$

$$= \dfrac{-14}{5}$$

g) $\left(\dfrac{D}{C}\right)(0) = \dfrac{D(0)}{C(0)}$

$$= \dfrac{(0)^2 + 1}{(7)(0)}$$

$$= \dfrac{1}{0}$$

which is undefined.

3.6 COMPOSITE FUNCTIONS

If $y = f(t) = t^2$ and $t = g(x) = x + 1$, we can write $y = h(x) = (x + 1)^2$ by substituting $(x + 1)$ for t in the equation $y = t^2$. Upon examining these functions, we see that the function h may be written as

$$h(x) = f(g(x))$$
$$= f(x + 1)$$
$$= (x + 1)^2$$

The function h is composed of the two functions f and g. The function h is called a composite function.

Definition:
The **composite function** f of g, denoted by $f{\circ}g$, is defined by
$$(f{\circ}g)(x) = f(g(x)),$$
where x is an element in the domain of g and $g(x)$ is an element in the domain of f.

> **Caution:**
> - The composite function $f \circ g$ is not the same as the product function $f \cdot g$.
> - In general, the composite function $f \circ g$ is not the same as the composite function $g \circ f$. The order in forming the composition of two functions is important.

EXERCISE 3.43

Let $A = \{1, 2, 3, 4\}$, $B = \{a, b, c, d\}$, $C = \{a, b, c\}$, and $D = \{-3, -2, 0, 4, 7\}$. Let f be the function defined from A into B as given in Figure 3.16, and let g be the function defined from C into D as given in Figure 3.17. If possible, form the function $g \circ f$.

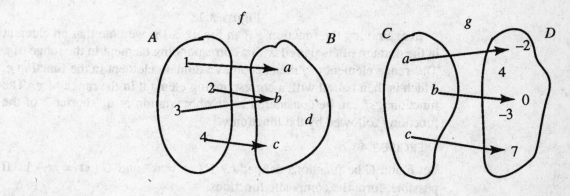

Figure 3.16   Figure 3.17

SOLUTION 3.43

$(g \circ f)(1) = g(f(1)) = g(a) = -2$

$(g \circ f)(2) = g(f(2)) = g(b) = 0$

$(g \circ f)(3) = g(f(3)) = g(b) = 0$

$(g \circ f)(4) = g(f(4)) = g(c) = 7$

The composite function $g \circ f$ is diagrammed in Figure 3.18.

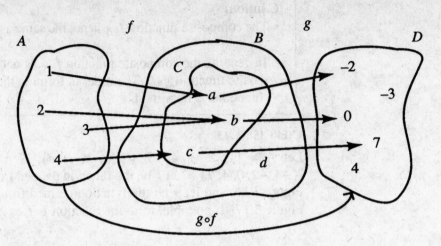

Figure 3.18

In examining the function $g \circ f$ in Figure 3.18, we note that an element in the domain of f is paired with a corresponding element in the range of f. This range element of f then becomes a domain element in the function g, which is then paired with a corresponding element in the range of g. The function $g \circ f$ can be considered as an abbreviation or a "shortcut" of the function f followed by the function g.

EXERCISE 3.44

Let F and G be functions defined by $F(x) = x^2$ and $G(x) = x - 1$. If possible, form the composite functions:

a) $F \circ G$

b) $G \circ F$

SOLUTION 3.44

a) $(F \circ G)(x) = F(G(x))$

$\qquad\qquad = F(x-1)$

$\qquad\qquad = (x-1)^2$

$D_{F \circ G} = (-\infty, +\infty)$

b) $(G \circ F)(x) = G(F(x))$

$\qquad\qquad = G(x^2)$

$\qquad\qquad = x^2 - 1$

$D_{G \circ F} = (-\infty, +\infty)$

EXERCISE 3.45

Let p and q be functions defined by $p(x) = \sqrt{x-1}$ and $q(x) = x^2$. If possible, form the composite functions:

a) $p \circ q$

b) $q \circ p$

SOLUTION 3.45

a) $(p \circ q)(x) = p(q(x))$

$$= p(x^2) \quad (x \text{ is any real number})$$

$$= \sqrt{x^2 - 1}$$

We determine that $x^2 - 1 \geq 0$ only if x belongs to $(-\infty, -1] \cup [1, +\infty)$. Therefore, $D_{p \circ q} = (-\infty, -1] \cup [1, +\infty)$.

b) $(q \circ p)(x) = q(p(x))$

$$= q(\sqrt{x-1}), \text{ if } x \geq 1$$

$$= (\sqrt{x-1})^2$$

$$= x - 1$$

Therefore, $D_{q \circ p} = [1, +\infty)$.

Caution:

In Exercise 3.44b, if you wait until the final step to determine the domain of the composite function, you may be inclined to state that the domain for the function $q \circ p$ is the set of all real numbers since $x - 1$ is defined for all real numbers. However, you must start with x in the domain of p and carry any restrictions throughout the exercise.

EXERCISE 3.46

Let s and r be two functions defined by $s(x) = \dfrac{1}{x}$ and $r(x) = 2x + 1$. If possible, form the composite functions:

a) $r \circ s$

b) $s \circ r$

SOLUTION 3.46

a) $(r \circ s)(x) = r(s(x))$

$$= r\left(\frac{1}{x}\right), \text{ if } x \neq 0$$

$$= 2\left(\frac{1}{x}\right) + 1$$

$$= \frac{2}{x} + 1$$

$$= \frac{2+x}{x}$$

$$D_{ros} = (-\infty, 0) \cup (0, +\infty).$$

b) $(s \circ r)(x) = s(r(x))$

$$= s(2x + 1) \text{ for all real values of } x$$

$$= \frac{1}{2x + 1}, \text{ provided that } x \neq \frac{-1}{2}$$

$$D_{s \circ r} = \left(-\infty, \frac{-1}{2}\right) \cup \left(\frac{-1}{2}, +\infty\right).$$

EXERCISE 3.47

Let A and B be two functions defined by $A(x) = \frac{1}{x-2}$ and $B(x) = x^3$.
If possible, form the composite functions:

a) $A \circ B$

b) $B \circ A$

SOLUTION 3.47

a) $(A \circ B)(x) = A(B(x))$

$$= A(x^3) \text{ for all } x \text{ in } R$$

$$= \frac{1}{x^3 - 2}, \text{ provided that } x \neq \sqrt[3]{2}$$

$$D_{A \circ B} = (-\infty, \sqrt[3]{2}) \cup (\sqrt[3]{2}, +\infty).$$

b) $(B \circ A)(x) = B(A(x))$

$$= B\left(\frac{1}{x-2}\right), \text{ provided that } x \neq 2$$

$$= \left(\frac{1}{x-2}\right)^3$$

$$= \frac{1}{(x-2)^3}$$

$$D_{B \circ A} = (-\infty, 2) \cup (2, +\infty).$$

3.7 INVERSE FUNCTIONS

Let f be a one-to-one function defined by $y = f(x)$. Then for each x
in the domain of f, there is exactly one y in the range of the function. But

because of the one-to-one nature of the function, for each *y* in the range of *f*, there is exactly one *x* in the domain. The correspondence from the range of *f* onto the domain of *f* is also a function. It is called the inverse function of *f* and is denoted by f^{-1}.

Definition:

Let *f* be a one-to-one function defined by $y = f(x)$. The **inverse** of *f*, denoted by f^{-1}, is a function such that

$$f^{-1}(f(x)) = x \text{ for every } x \text{ in the domain of } f, \text{ and}$$

$$f(f^{-1}(x)) = x \text{ for every } x \text{ in the domain of } f^{-1}$$

Caution:

The symbol f^{-1} does *not* mean $\dfrac{1}{f}$. The −1 is *not* an exponent. The symbol f^{-1} means inverse of a function as a function.

In Figure 3.19, we illustrate the relationship between a function and its inverse function.

Facts:

• The domain of a function is equal to the range of its inverse function.

• The range of a function is equal to the domain of its inverse function.

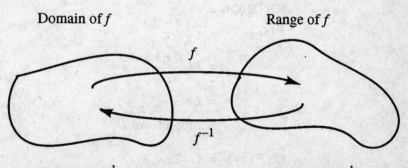

Figure 3.19

From the above definition, it should be clear that if two functions are inverses of each other, then each function undoes what the other function does. Also, we note that when forming the composition of a function and its inverse function, we get the Identity function.

EXERCISE 3.48

Let f be the function defined by $y = f(x) = 2x$, and let g be the function defined by $y = g(x) = \dfrac{x}{2}$. Verify that the function g is the inverse of f.

SOLUTION 3.48

$$f(g(x)) = f\left(\frac{x}{2}\right)$$

$$= 2\left(\frac{x}{2}\right)$$

$$= x \text{ for all } x \text{ in } D_g$$

$$g(f(x)) = g(2x)$$

$$= \frac{2x}{2}$$

$$= x \text{ for all } x \text{ in } D_f$$

Therefore, $g = f^{-1}$

EXERCISE 3.49

Let F be the function defined by $F(x) = 2x - 3$, and let G be the function defined by $y = G(x) = \dfrac{x+3}{2}$. Verify that the function G is the inverse of F.

SOLUTION 3.49

$$F(G(x)) = F\left(\frac{x+3}{2}\right)$$

$$= 2\left(\frac{x+3}{2}\right) - 3$$

$$= x + 3 - 3$$

$$= x \text{ for all } x \text{ in } D_G$$

$$G(F(x)) = G(2x - 3)$$

$$= \frac{(2x - 3) + 3}{2}$$

$$= \frac{2x - 3 + 3}{2}$$

$$= \frac{2x}{2}$$

$$= x \text{ for all } x \text{ in } D_F$$

Therefore, $G = F^{-1}$.

Consider the one-to-one function f defined by $y = f(x) = 2x + 1$. As a set of ordered pairs, f contains $(-2, -3)$, $(-1, -1)$, $(0, 1)$, $(1, 3)$, and $(2, 5)$. The graph of f is a line passing through the points whose coordinates are these ordered pairs. The inverse function f^{-1}, then, contains the ordered pairs $(-3, -2)$, $(-1, -1)$, $(1, 0)$, $(3, 1)$, and $(5, 2)$. The graph of f^{-1} is also a line. The graphs of f and f^{-1} are given in Figure 3.20. The graph of f^{-1} is

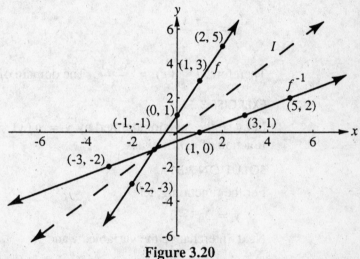

Figure 3.20

the reflection of the graph of f about the line $y = x$, which is the graph of the Identity function.

From an examination of the graphs in Figure 3.20, we observe that we can obtain the graph of f^{-1} by interchanging the roles of x and y in each of the ordered pairs in f. We will use this procedure to form the inverse of a function, if its inverse exists.

Procedure for Finding the Inverse of a Function

To find the inverse of the function f defined by $y = f(x)$:

Step 1: Write $y = f(x)$.

Step 2: Interchange the role of the variables x and y.

Step 3: Solve for y in the equation formed in Step 2.

Step 4: If the equation in Step 3 defines a function, write $y = f^{-1}(x)$.

EXERCISE 3.50

Given the function f defined by $y = f(x) = 3x - 1$, determine its inverse function, f^{-1}, if it exists.

SOLUTION 3.50

For the function f, let

$$y = 3x - 1$$

Interchange the variables x and y, obtaining

$$x = 3y - 1$$

Solve for y, obtaining

$$x = 3y - 1$$
$$3y - 1 = x$$
$$3y = x + 1$$

$$y = \frac{x + 1}{3}$$

Therefore, $f^{-1}(x) = \dfrac{x + 1}{3}$. The domain of f^{-1} is $(-\infty, +\infty)$.

EXERCISE 3.51

Let g be the function defined by $y = g(x) = 2x^2 - 1$. Form the inverse function of g, if it exists.

SOLUTION 3.51

For the function g, let

$$y = 2x^2 - 1$$

Next, interchange the variables x and y.

$$x = 2y^2 - 1$$

Solve for y.

$$x = 2y^2 - 1$$
$$2y^2 - 1 = x$$
$$2y^2 = x + 1$$

$$y^2 = \frac{x + 1}{2}$$

$$y = \pm\sqrt{\frac{x + 1}{2}}$$

We now determine that y does not define a function. For instance, if $x = 1$, then $y = \pm 1$. Therefore, the function g *does not* have an inverse as a function.

EXERCISE 3.52

Let G be the function defined by $y = G(x) = x^2$ for $x \geq 0$. If G has an inverse as a function, form it.

SOLUTION 3.52

Step 1: $y = x^2$ for $x \geq 0$

Step 2:   Interchange the variables x and y.

                          $x = y^2$

Step 3:   Solve for y.

                          $x = y^2$

                          $y^2 = x$

                          $y = \pm\sqrt{x}$

                          Since $x \geq 0$, then $y = x^2 \geq 0$. Reject $y = -\sqrt{x}$.

Step 4: $y = G(x) = \sqrt{x}$ defines the inverse function of G.

*I*n this chapter, some special functions were introduced. The algebra of functions and composition of functions were discussed. The chapter concluded with an examination of the inverse of a function.

SUPPLEMENTAL EXERCISES

In Exercises 1–4, determine if the given function is increasing, decreasing, or neither on the indicated interval.

1. $f(x) = 2 - x$; $[-3, 4]$

2. $g(x) = |x|$; $[-6, -1]$

3. $h(x) = \dfrac{-1}{x}$; $[-4.1, -1]$

4. $j(x) = x^2 - 1$; $[-2, 3]$

In Exercises 5–8, determine if the given function is even, odd, or neither.

5. $F(x) = 2x - 1$

6. $G(x) = x^3 - 3x$

7. $H(x) = -2x^2 + 2$

8. $J(x) = 2.6x$

In Exercises 9–10, determine whether the graph of the given function is symmetric with respect to the y-axis, the origin, or neither.

9. $y = p(x) = 4x - x^3$

10. $y = q(x) = x^2 - 2x + 7$

In Exercises 11–16, let $f(x) = 3x - 4$, $g(x) = x^2 - 1$, $h(x) = \sqrt{x - 5}$, and $j(x) = \dfrac{2}{3x}$. Form each of the following functions and state its domain.

11. $f + h$

12. $g \cdot j$

13. $\dfrac{f}{h}$

14. $f \circ g$

15. $g \circ h$

16. $j \circ f$

17. Let $R(x) = x - 3$ and $S(x) = 3 + x$. Is $S = R^{-1}$?

18. Let $T(x) = 3x - 1$ and $W(x) = \dfrac{1}{3x - 1}$. Is $W = T^{-1}$?

19. Let $y = r(x) = 2 - 3x$. Form r^{-1}, if it exists, and determine its domain.

20. Let $y = u(x) = x^2 + 3$. Form u^{-1}, if it exists, and determine its domain.

ANSWERS TO SUPPLEMENTAL EXERCISES

1. Decreasing

2. Decreasing

3. Increasing

4. Neither

5. Neither

6. Odd

7. Even

8. Odd

9. Origin

10. Neither

11. $(f + h)(x) = 3x - 4 + \sqrt{x - 5}$; domain $= [5, +\infty)$

12. $(g \cdot j)(x) = \dfrac{2(x^2 - 1)}{3x}$; domain $= (-\infty, 0) \cup (0, +\infty)$

13. $\left(\dfrac{f}{h}\right)(x) = \dfrac{3x-4}{\sqrt{x-5}}$; domain $= (5, +\infty)$

14. $(f \circ g)(x) = 3x^2 - 7$; domain $= (-\infty, +\infty)$

15. $(g \circ h)(x) = x - 6$; domain $= [5, +\infty)$

16. $(j \circ f)(x) = \dfrac{2}{9x - 12}$; domain $= \left(-\infty, \dfrac{4}{3}\right) \cup \left(\dfrac{4}{3}, +\infty\right)$

17. Yes

18. No

19. $r^{-1}(x) = \dfrac{2-x}{3}$; domain $= (-\infty, +\infty)$

20. u^{-1} does not exist.

4

Techniques for Graphing Functions

*F*rom an accurate graph, you can generally discuss the behavior and the characteristics of a function. In this chapter, we examine some techniques that will enable you to sketch the graph of a function without having to plot an unnecessary number of points. We shall examine the graphs of functions that are variations of what are considered to be some of the most basic functions that were discussed in Chapter 3.

4.1 VERTICAL SHIFTS

Suppose that we have the graph, in the xy-plane, of a function f defined by $y = f(x)$. (See Figure 4.1.)

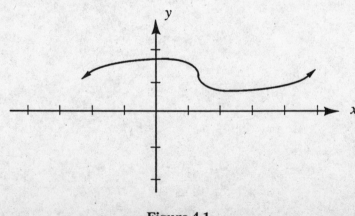

Figure 4.1
The Graph of the Function f

Now, suppose that we wish to construct the graph, also in the xy-plane, of a function g defined by $y = g(x) = f(x) + 2$. Clearly, the function g is the sum of the function f and the Constant-Valued function with specification $y = 2$. Hence, $g(a) = f(a) + 2$ for *every* a in the domain of f. The graph of g, then, would simply be the graph of f raised 2 units in the xy-plane. In a similar manner, the function h defined by $h(x) = f(x) - 3$ would have a graph in the xy-plane that is the graph of f lowered 3 units. (See Figure 4.2.)

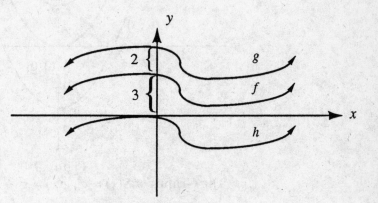

Figure 4.2
Graphs of the Functions $f, g,$ and h

In chapter 3, we considered the Squaring function, defined by $y = S(x) = x^2$, and its graph. (See Figure 4.3.)

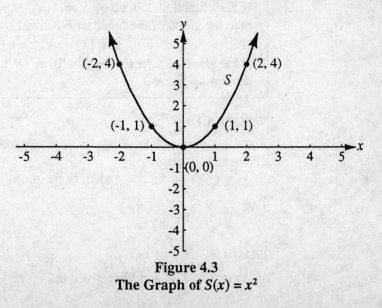

Figure 4.3
The Graph of $S(x) = x^2$

The graph of $g(x) = x^2 + 2$, then, would be the graph of S raised 2 units, and the graph of $h(x) = x^2 - 3$ would be the graph of S lowered 3 units. (See Figure 4.4.)

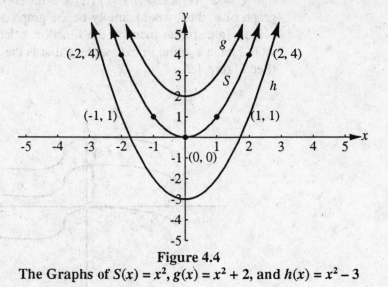

Figure 4.4
The Graphs of $S(x) = x^2$, $g(x) = x^2 + 2$, and $h(x) = x^2 - 3$

In general, we have the following rule relative to vertical shifts of graphs.

Rule:

If f is a function defined by $y = f(x)$ and k is a *nonzero* constant, then the graph of the function F defined by
$$F(x) = f(x) + k$$
is the graph of f **shifted vertically** by $|k|$ units, upward if $k > 0$ and downward if $k < 0$.

EXERCISE 4.1

On the same set of coordinate axes, graph and label each of the functions defined as follows:

a) $y = I(x) = x$ (Identity function)

b) $y = g(x) = x + 2$

c) $y = h(x) = x - 5$

SOLUTION 4.1

The graphs of I, g, and h are given in Figure 4.5.

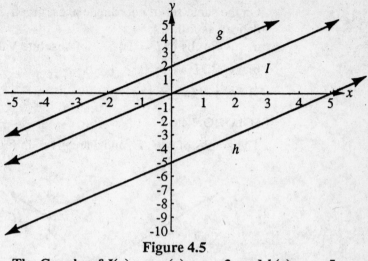

Figure 4.5

The Graphs of $I(x) = x$, $g(x) = x + 2$, and $h(x) = x - 5$

EXERCISE 4.2

On the same set of coordinate axes, graph and label each of the functions defined as follows:

a) $y = S(x) = x^2$ (Squaring function)

b) $y = j(x) = x^2 + 3$

c) $p(x) = x^2 - 1$

SOLUTION 4.2

The graphs of S, j, and p are given in Figure 4.6.

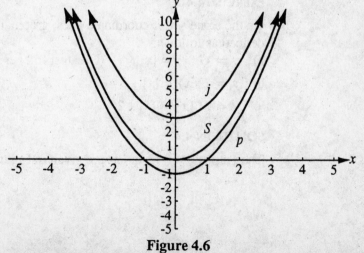

Figure 4.6

The Graphs of $S(x) = x^2$, $j(x) = x^2 + 3$, and $p(x) = x^2 - 1$

EXERCISE 4.3

On the same set of coordinate axes, graph and label each of the functions defined as follows:

a)　$y = \textbf{Abs}\,(x) = |x|$　　　(Absolute Value function)

b)　$y = H(x) = |x| + 4$

c)　$y = J(x) = |x| - 4$

SOLUTION 4.3

The graphs of **Abs**, H, and J are given in Figure 4.7.

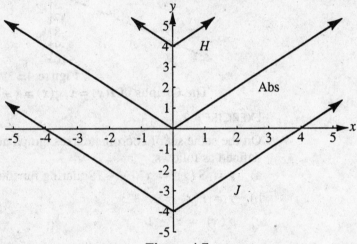

Figure 4.7

The Graphs of Abs(x) = | x |, H(x) = | x | + 4, and J(x) = | x | − 4

EXERCISE 4.4

On the same set of coordinate axes, graph and label each of the functions defined as follows:

a)　$y = G(x) = [x]$　　　(Greatest Integer function)

b)　$y = F(x) = [x] + 3$

c)　$y = M(x) = [x] - 2$

SOLUTION 4.4

The graphs of G, F, and M are given in Figure 4.8.

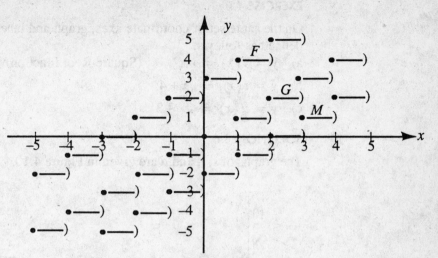

Figure 4.8
The Graphs of $G(x) = [x]$, $F(x) = [x] + 3$, and $M(x) = [x] - 2$

EXERCISE 4.5

On the same set of coordinate axes, graph and label each of the functions defined as follows:

a) $y = C(x) = x^3$ (Cubing function)

b) $y = f(x) = x^3 + 4$

c) $y = h(x) = x^3 - 2$

SOLUTION 4.5

The graphs of C, f, and h are given in Figure 4.9.

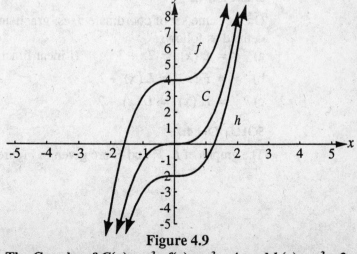

Figure 4.9
The Graphs of $C(x) = x^3$, $f(x) = x^3 + 4$, and $h(x) = x^3 - 2$

EXERCISE 4.6

On the same set of coordinate axes, graph and label each of the functions defined as follows:

a) $y = s(x) = \sqrt{x}$ (Square Root function)

b) $y = t(x) = \sqrt{x} + 4$

c) $y = u(x) = \sqrt{x} - 3$

SOLUTION 4.6

The graphs of s, t, and u are given in Figure 4.10.

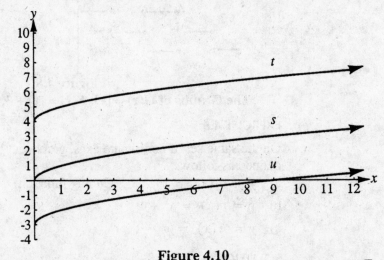

Figure 4.10

The Graphs of $s(x) = \sqrt{x}$, $t(x) = \sqrt{x} + 4$, and $u(x) = \sqrt{x} - 3$

EXERCISE 4.7

On the same set of coordinate axes, graph and label each of the functions defined as follows:

a) $y = L(x) = 2x + 3$ (Linear function)

b) $y = F(x) = L(x) + 2$

c) $y = H(x) = L(x) - 2$

SOLUTION 4.7

The graphs of L, F, and H are given in Figure 4.11.

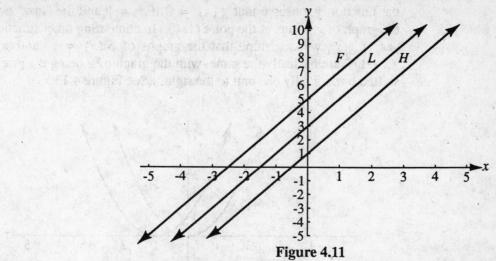

Figure 4.11

The Graphs of $L(x) = 2x + 3$, $F(x) = L(x) + 2$, and $H(x) = L(x) - 2$

4.2 HORIZONTAL SHIFTS

Consider the function S defined by $y = S(x) = x^2$. We observe that $S(x) = 0$ if $x = 0$ and that the "low" point on the graph of S occurs at the origin of the xy-plane. (See Figure 4.12.)

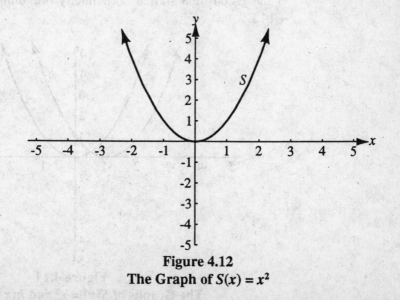

Figure 4.12

The Graph of $S(x) = x^2$

Now, consider the function g such that $y = g(x) = (x-1)^2$. For the function g, observe that $g(x) = 0$ if $x = 1$ and the "low" point on the graph of g occurs at the point $(1, 0)$. In computing other function values of $g(x)$, we determine that the graphs of $S(x) = x^2$ and $g(x) = (x-1)^2$ are basically the same, with the graph of g being the graph of S shifted horizontally one unit to the right. (See Figure 4.13.)

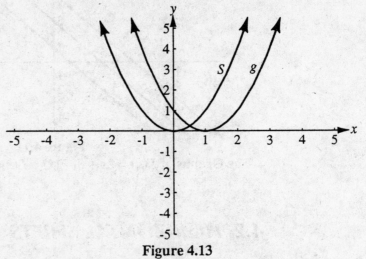

Figure 4.13
The Graphs of $S(x) = x^2$ and $g(x) = (x-1)^2$

In a similar manner, the graph of h defined by $y = h(x) = (x+1)^2$ is the graph of S shifted horizontally one unit to the left. (See Figure 4.14.)

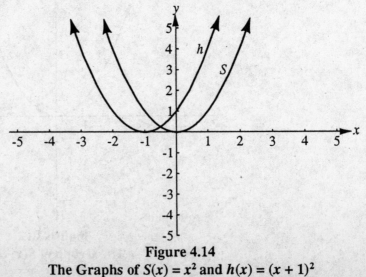

Figure 4.14
The Graphs of $S(x) = x^2$ and $h(x) = (x + 1)^2$

In general, we have the following rule relative to horizontal shifts of graphs.

Rule:

If f is a function defined by $y = f(x)$ and h is a *positive* constant, then the graph of the function F defined by
$$F(x) = f(x - h)$$
is the graph of f **shifted horizontally** by h units to the **right**. The graph of the function G defined by
$$G(x) = f(x + h)$$
is the graph of f **shifted horizontally** by h units to the **left**.

EXERCISE 4.8

On the same set of coordinate axes, graph and label each of the functions defined as follows:

a) $y = S(x) = x^2$

b) $y = F(x) = (x - 2)^2$

c) $y = H(x) = (x + 3)^2$

SOLUTION 4.8

The graphs of S, F, and H are given in Figure 4.15.

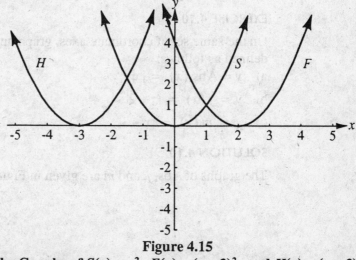

Figure 4.15

The Graphs of $S(x) = x^2$, $F(x) = (x - 2)^2$, and $H(x) = (x + 3)^2$

EXERCISE 4.9

On the same set of coordinate axes, graph and label each of the functions defined as follows:

a) $y = C(x) = x^3$

b) $y = f(x) = (x-4)^3$

c) $y = g(x) = (x+3)^3$

SOLUTION 4.9

The graphs of C, f, and g are given in Figure 4.16.

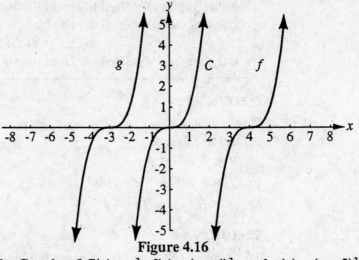

Figure 4.16

The Graphs of $C(x) = x^3$, $f(x) = (x-4)^3$, and $g(x) = (x+3)^3$

EXERCISE 4.10

On the same set of coordinate axes, graph and label each of the functions defined as follows:

a) $y = \text{Abs}(x) = |x|$

b) $y = j(x) = |x+2|$

c) $y = m(x) = |x-3|$

SOLUTION 4.10

The graphs of **Abs**, j, and m are given in Figure 4.17.

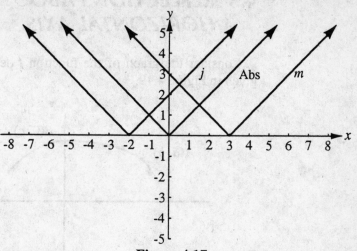

Figure 4.17
The Graphs of Abs(x) = |x|, j(x) = |x + 2|, and m(x) = |x − 3|

EXERCISE 4.11

On the same set of coordinate axes, graph and label each of the functions defined as follows:

a) $y = s(x) = \sqrt{x}$

b) $y = T(x) = \sqrt{x+2}$

c) $y = V(x) = \sqrt{x-4}$

SOLUTION 4.11

The graphs of *s*, *T*, and *V* are given in Figure 4.18.

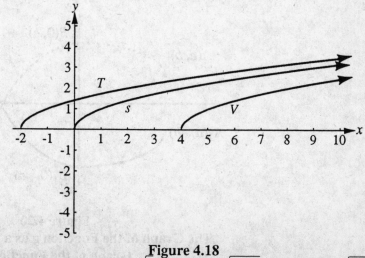

Figure 4.18
The Graphs of $s(x) = \sqrt{x}$, $T(x) = \sqrt{x+2}$, and $V(x) = \sqrt{x-4}$

4.3 REFLECTIONS ABOUT THE HORIZONTAL AXIS

Consider the graph of the function f defined by $y = f(x)$, which is given in Figure 4.19.

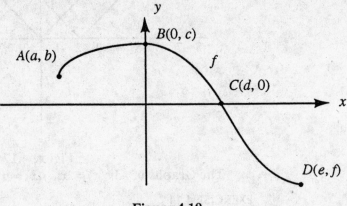

Figure 4.19
The Graph of the Function f

Now, consider the function g defined by $g(x) = -f(x)$. Its graph can be obtained quite readily. For each a in the domain of f, $g(a) = -f(a)$. That is, if the point $P(x, y)$ is on the graph of f, then the point $P_1(x, -y)$ will be on the graph of g. For the points $A(a, b)$, $B(0, c)$, $C(d, 0)$, and $D(e, f)$ on the graph of f, there would be the points $A_1(a, -b)$, $B_1(0, -c)$, $C_1(d, 0)$, and $D_1(e, -f)$ on the graph of g. (See Figure 4.20.)

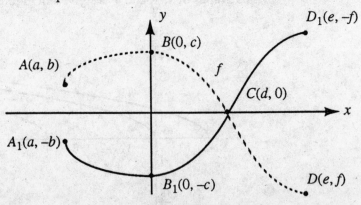

Figure 4.20
**The Graph of the Function g as a Reflection of the
Graph of the Function f**

Observe that the graph of g is the mirror image of the graph of f about the x-axis. In general, if f and g are two functions such that $g(x) = -f(x)$ for all x in the domain of f, then the graph of g is the mirror image of the graph of f. The graph of g is called a **reflection** of the graph of f about the horizontal axis.

EXERCISE 4.12

Graph the function g defined by $g(x) = -x^2$.

SOLUTION 4.12

The function g may be defined by $g(x) = -S(x)$ where S is the Squaring function defined by $y = S(x) = x^2$ for all real values of x. The graph of g, then, is the reflection of the graph of f about the x-axis. (See Figure 4.21.)

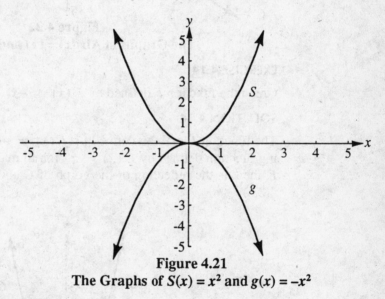

Figure 4.21
The Graphs of $S(x) = x^2$ and $g(x) = -x^2$

EXERCISE 4.13

Graph the function h defined by $h(x) = -|x|$.

SOLUTION 4.13

The function h may be defined by $h(x) = -\text{Abs}(x)$, where $\text{Abs}(x) = |x|$ for all real values of x. The graph of h, then, is the reflection of the graph of **Abs** about the x-axis. (See Figure 4.22.)

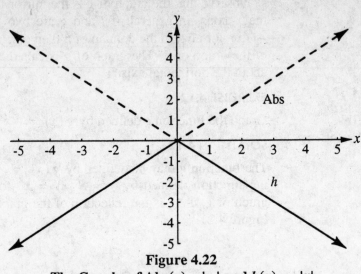

Figure 4.22
The Graphs of Abs(x) = |x| and $h(x)$ = –|x|

EXERCISE 4.14

Graph the function F defined by $F(x) = -x^3$.

SOLUTION 4.14

The function F may be defined by $F(x) = -C(x)$ where C is the Cubing function defined by $C(x) = x^3$ for all real values of x. The graph of F, then, is the reflection of the graph of C about the x-axis. (See Figure 4.23.)

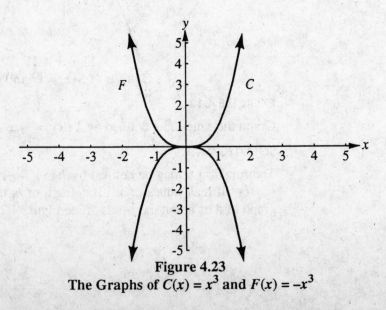

Figure 4.23
The Graphs of $C(x) = x^3$ and $F(x) = -x^3$

EXERCISE 4.15

Graph the function H defined by $H(x) = -\sqrt{x}$.

SOLUTION 4.15

The function H may be defined by $H(x) = -s(x)$, where s is the Square Root function defined by $s(x) = \sqrt{x}$ for all $x \geq 0$. The graph of H, then, is the reflection of the graph of s about the x-axis. (See Figure 4.24.)

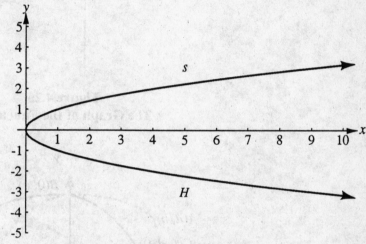

Figure 4.24
The Graphs of $s(x) = \sqrt{x}$ and $H(x) = -\sqrt{x}$

4.4 REFLECTIONS ABOUT THE VERTICAL AXIS

Consider the graph of the function f defined by $y = f(x)$, which is given in Figure 4.25. Now, consider the function g defined by $g(x) = f(-x)$. Its graph can be obtained quite readily. For each a in the domain of f, $g(a) = f(-a)$. That is, if the point $P(x, y)$ is on the graph of f, then the point $P_1(-x, y)$ will be on the graph of g. For the points $A(a, b)$, $B(0, c)$, $C(d, 0)$, and $D(e, f)$ on the graph of f, there would be the points $A_1(-a, b)$, $B(0, c)$, $C_1(-d, 0)$, and $D_1(-e, f)$ on the graph of g. (See Figure 4.26.)

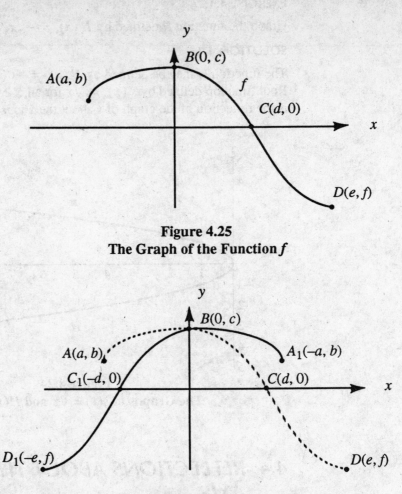

Figure 4.25
The Graph of the Function f

Figure 4.26
The Graph of the Function g as a Reflection of the
Graph of the Function f

Observe that the graph of g is the mirror image of the graph of f about the y-axis. In general, if f and g are two functions such that $g(x) = f(-x)$ for all x in the domain of f, then the graph of g is the mirror image of the graph of f. The graph of g is called a **reflection** of the graph of f about the vertical axis.

EXERCISE 4.16

Graph the function F defined by $F(x) = (-x)^3$.

SOLUTION 4.16

The function F may be defined by $F(x) = C(-x)$ where C is the Cubing function defined by $C(x) = x^3$ for all real values of x. The graph of

F, then, is the reflection of the graph of *C* about the *y*-axis. (See Figure 4.27.)

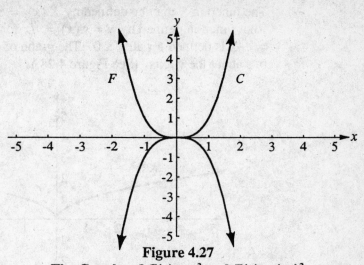

Figure 4.27
The Graphs of $C(x) = x^3$ and $F(x) = (-x)^3$

Reconsider $F(x) = (-x)^3$. Since $(-x)^3 = (-1)^3 (x)^3 = -(x)^3$, the graph of *F* is also a reflection of the graph of *C* about the *x*-axis. Verify this.

EXERCISE 4.17

What is the graph of $H(x) = (-x)^2$?

SOLUTION 4.17

Since the Squaring function, *S*, is defined by $y = S(x) = x^2$, then $H(x) = S(-x)$. Hence, the graph of *H* is a reflection of the graph of *S* about the *y*-axis. But, $(-x)^2 = (x)^2$ for all real values of *x*. Therefore, the graph of *H* is the graph of *S*. Recall that the graph of *S* is symmetric about the *y*-axis.

EXERCISE 4.18

What is the graph of $J(x) = |-x|$?

SOLUTION 4.18

Since the Absolute Value function, **Abs**, is defined by $y = \text{Abs}(x) = |x|$, then $J(x) = \text{Abs}(-x)$. Hence, the graph of *J* is a reflection of the graph of **Abs** about the *y*-axis. But, $|-x| = |x|$ for all real values of *x*. Therefore, the graph of *J* is the graph of **Abs**. Recall that the graph of **Abs** is symmetric about the *y*-axis.

EXERCISE 4.19

Graph the function *K* defined by $K(x) = \sqrt{-x}$.

SOLUTION 4.19

The function K may be defined by $K(x) = s(-x)$ where s is the Square Root function defined by $y = s(x) = \sqrt{x}$ for all $x \geq 0$. Therefore, $K(x) = \sqrt{-x}$ is defined for all $x \leq 0$. The graph of K is a reflection of the graph of s about the y-axis. (See Figure 4.28.)

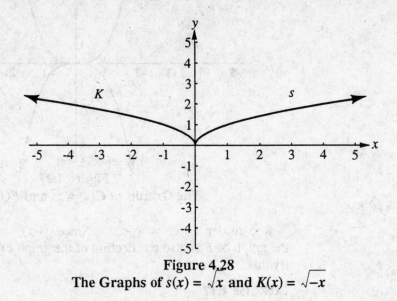

Figure 4.28
The Graphs of $s(x) = \sqrt{x}$ and $K(x) = \sqrt{-x}$

4.5 SCALAR MULTIPLES OF FUNCTIONS

Consider the graph of the function f defined by $y = f(x)$, which is given in Figure 4.29. The graph of the function g defined by $g(x) = 2f(x)$ can be obtained quite readily. For each a in the domain of f, $g(a) = 2f(a)$. That is, if the point $P(x, y)$ is on the graph of f, then the point $P_1(x, 2y)$ will be on the graph of g. For the points $A(a, b)$, $B(c, 0)$, $C(d, e)$, $D(0, 0)$, $E(f, g)$, $F(h, 0)$, and $G(i, j)$ on the graph of f, there will be the points $A_1(a, 2b)$, $B(c, 0)$, $C_1(d, 2e)$, $D(0, 0)$, $E_1(f, 2g)$, $F(h, 0)$, and $G_1(i, 2j)$ on the graph of g. (See Figure 4.30.)

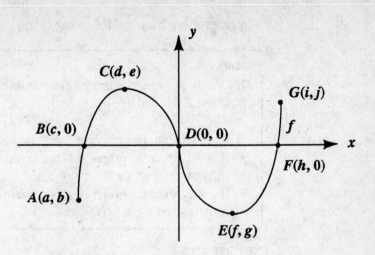

Figure 4.29
The Graph of the Function *f*

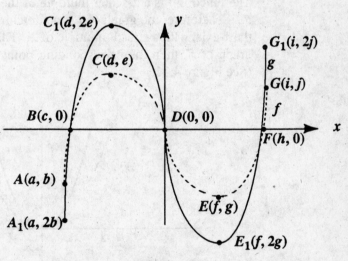

Figure 4.30
The Graphs of the Functions *f* and *g*

For each point on the graph of *f*, there is a corresponding point on the graph of *g* with a *y*-coordinate that is twice the *y*-coordinate of the point with the same *x*-coordinate on the graph of *f*.

In general, we have the following rule.

> **Rule:**
> If f is a function and a is a *positive* constant, then the graph of the function F defined by
> $$F(x) = af(x)$$
> is obtained from the graph of f as follows:
> - If $a > 1$, then the graph of F is a **stretching** of the graph of f vertically away from the x-axis.
> - If $a < 1$, then the graph of F is a **shrinking** of the graph of f vertically toward the x-axis.

EXERCISE 4.20

Graph the function f defined by $f(x) = 2x^2$.

SOLUTION 4.20

The function f is a 2-scalar multiple of the function S defined by $S(x) = x^2$. Therefore, the graph of f is a stretching of the graph of S away from the x-axis, with a stretch multiple of 2. That is, if the point (a, b) is on the graph of S, then the corresponding point $(a, 2b)$ is on the graph of f. (See Figure 4.31.)

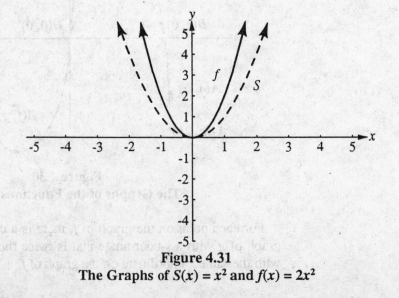

Figure 4.31
The Graphs of $S(x) = x^2$ and $f(x) = 2x^2$

EXERCISE 4.21

Graph the function h defined by $h(x) = 3|x|$.

SOLUTION 4.21

The function h is a 3-scalar multiple of the function **Abs** defined by **Abs** $(x) = |x|$. Therefore, the graph of h is a stretching of the graph of **Abs** away from the x-axis, with a stretch multiple of 3. That is, if the point (a, b) is on the graph of **Abs**, then the corresponding point $(a, 3a)$ is on the graph of h. (See Figure 4.32.)

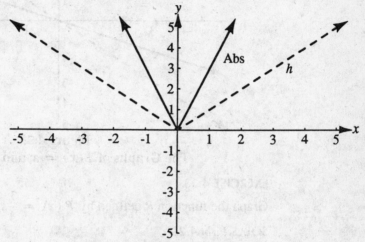

Figure 4.32
The Graphs of Abs(x) = |x| and h(x) = 3|x|

EXERCISE 4.22

Graph the function P defined by $P(x) = \frac{1}{2}x$.

SOLUTION 4.22

The function P is a $\frac{1}{2}$-scalar multiple of the function I defined by $I(x) = x$. Therefore, the graph of P is a shrinking of the graph of I toward the x-axis, with a shrink multiple of $\frac{1}{2}$. That is, if the point (a, b) is on the graph of I, then the corresponding point $\left(a, \frac{1}{2}b\right)$ is on the graph of P. (See Figure 4.33.)

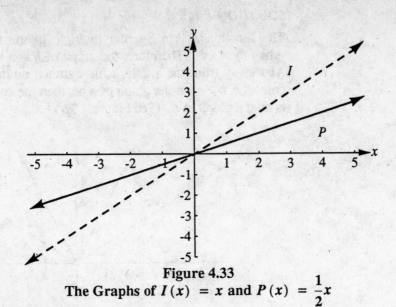

Figure 4.33
The Graphs of $I(x) = x$ and $P(x) = \dfrac{1}{2}x$

EXERCISE 4.23

Graph the function R defined by $R(x) = \dfrac{1}{3}x^3$.

SOLUTION 4.23

The function R is a $\dfrac{1}{3}$-scalar multiple of the function C defined by $C(x) = x^3$. Therefore, the graph of R is a shrinking of the graph of C toward the x-axis, with a shrink multiple of $\dfrac{1}{3}$. That is, if the point (a, b) is on the graph of C, then the corresponding point $\left(a, \dfrac{1}{3}b\right)$ is on the graph of R. (See Figure 4.34.)

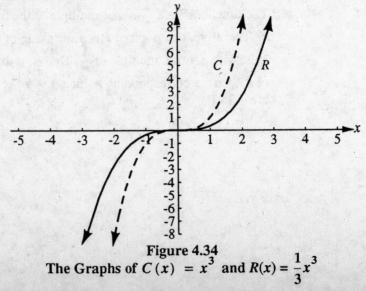

Figure 4.34
The Graphs of $C(x) = x^3$ and $R(x) = \dfrac{1}{3}x^3$

4.6 MORE ON GRAPHING OF FUNCTIONS

In the previous sections of this chapter, we examined various techniques for graphing functions. In each section, we considered only one of the following: a vertical shift; a horizontal shift; a reflection about either the horizontal or the vertical axis; and the graph of a scalar multiple of a function. In this section, we combine two or more of these techniques in the same exercise.

EXERCISE 4.24

Graph the function P defined by $P(x) = (x+1)^2 - 2$.

SOLUTION 4.24

The function P is the sum of the two functions f and g defined by $f(x) = (x+1)^2$ and $g(x) = -2$ for all real values of x. But, the function f may also be defined by $f(x) = S(x+1)$ where S is the Squaring function. Hence, the graph of P is the graph of S shifted horizontally one unit to the left (obtaining the graph of f) and then shifted vertically two units downward. (See Figure 4.35.)

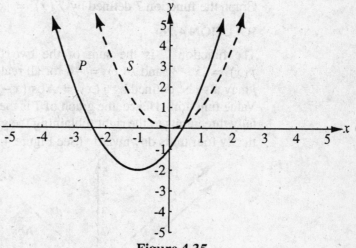

Figure 4.35
The Graphs of $S(x) = x^2$ and $P(x) = (x+1)^2 - 2$

EXERCISE 4.25

Graph the function Q defined by $y = Q(x) = (x-1)^3 + 2$.

SOLUTION 4.25

The function Q is the sum of the two functions p and q defined by $p(x) = (x-1)^3$ and $q(x) = 2$ for all real values of x. But, the function p may also be defined by $p(x) = C(x-1)$ where C is the Cubing function. Hence, the graph of Q is the graph of C shifted horizontally one unit to the right (obtaining the graph of p) and then shifted vertically two units upward. (See Figure 4.36.)

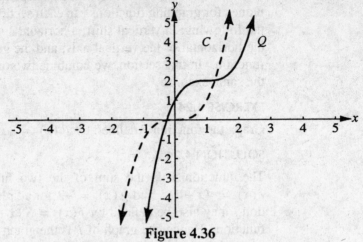

Figure 4.36

The Graphs of $C(x) = x^3$ and $Q(x) = (x - 1)^3 + 2$

EXERCISE 4.26

Graph the function T defined by $T(x) = |x - 3| - 4$.

SOLUTION 4.26

The function T is the sum of the two functions j and k defined by $j(x) = |x - 3|$ and $k(x) = -4$ for all real values of x. But, the function j may also be defined by $j(x) = \text{Abs}(x - 3)$ where **Abs** is the Absolute Value function. Hence, the graph of T is the graph of **Abs** shifted horizontally three units to the right (obtaining the graph of j) and then shifted vertically four units downward. (See Figure 4.37.)

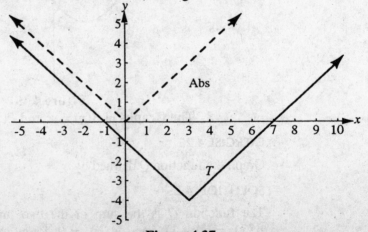

Figure 4.37

The Graphs of Abs(x) = |x| and $T(x) = |x - 3| - 4$

EXERCISE 4.27

Graph the function M defined by $M(x) = 2(x+2)^2$.

SOLUTION 4.27

To graph the function M, we start with the graph of the Squaring function, shift it horizontally two units to the left, and then stretch the resulting graph, using a stretch factor of 2. (See Figure 4.38.)

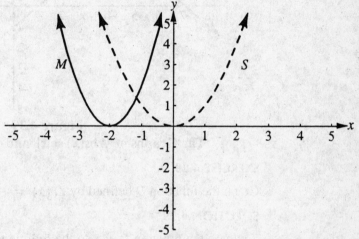

Figure 4.38

The Graphs of $S(x) = x^2$ and $M(x) = 2(x+2)^2$

EXERCISE 4.28

Graph the function E defined by $y = E(x) = -|x+4| + 3$.

SOLUTION 4.28

To graph the function E, we do the following:

Step 1:

Start with the graph of **Abs**.

Step 2:

Shift the graph of **Abs** horizontally four units to the left (obtaining the graph of $y = |x+4|$).

Step 3:

Reflect the graph of $y = |x+4|$ about the x-axis (obtaining the graph of $y = -|x+4|$).

Step 4:

Shift the graph of $y = -|x+4|$ vertically three units upward (obtaining the graph of E). (See Figure 4.39.)

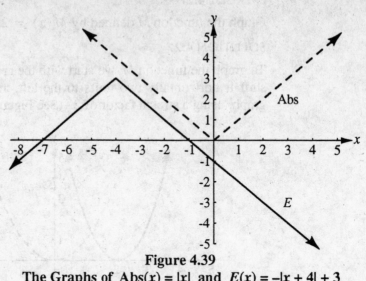

Figure 4.39
The Graphs of Abs(*x*) = l*x*l and *E*(*x*) = −l*x* + 4l + 3

EXERCISE 4.29

Graph the function *D* defined by $D(x) = -2(x-3)^2 + 4$.

SOLUTION 4.29

To graph the function *D*, we do the following:

Step 1:

Start with the graph of the Squaring function, *S*.

Step 2:

Shift the graph of *S* horizontally three units to the right (obtaining the graph of $y = (x-3)^2$).

Step 3:

Stretch the graph of $y = (x-3)^2$, using a stretch factor of 2 (obtaining the graph of $y = 2(x-3)^2$).

Step 4:

Reflect the graph of $y = 2(x-3)^2$ about the x-axis (obtaining the graph of $y = -2(x-3)^2$).

Step 5:

Shift the graph of $y = -2(x-3)^2$ vertically four units upward (obtaining the graph of *D*). (See Figure 4.40.)

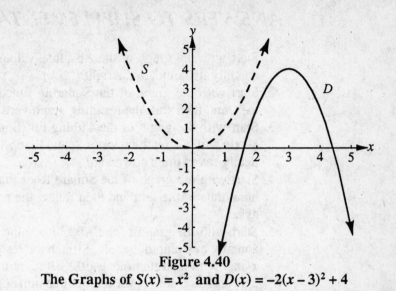

Figure 4.40

The Graphs of $S(x) = x^2$ and $D(x) = -2(x-3)^2 + 4$

In this chapter, we considered the graphing of functions using various techniques. Starting with the graph of a standard function, the graph of the new function was obtained by using a horizontal shift, a vertical shift, a reflection about the x-axis, or a combination of these techniques.

SUPPLEMENTAL EXERCISES

For each of the following exercises, start with the graph of a standard function and indicate how you would get the graph of the defined function.

1. $y = A(x) = |x - 5|$

2. $y = B(x) = -x^2 + 6$

3. $y = D(x) = 2x^3 - 7$

4. $y = E(x) = -\sqrt{x + 9}$

5. $y = F(x) = -2|x + 1| + 3$

6. $y = H(x) = \dfrac{1}{2}\sqrt{x} + 2$

7. $y = J(x) = 0.6\sqrt[3]{x + 4} - 5$

8. $y = K(x) = -[x + 1.5] - 5$

9. $y = M(x) = 3 - 4(x + 1)^3$

10. $y = N(x) = -4(x - 2)^2 + 7$

ANSWERS TO SUPPLEMENTAL EXERCISES

1. Start with the graph of the Absolute Value function and shift it horizontally five units to the right.

2. Start with the graph of the Squaring function, reflect it about the x-axis, and then shift the resulting graph vertically six units upward.

3. Start with the graph of the Cubing function, stretch it away from the x-axis with a stretch factor of 2, and then shift the resulting graph vertically seven units downward.

4. Start with the graph of the Square Root function, shift it horizontally nine units to the left, and then reflect the resulting graph about the x-axis.

5. Start with the graph of the Absolute Value function and shift it horizontally one unit to the left. Stretch the resulting graph away from the x-axis with a stretch factor of 2. Reflect the new graph about the x-axis and then shift it vertically three units upward.

6. Start with the graph of the Square Root function, shrink it toward the x-axis with a shrink factor of $\frac{1}{2}$, and then shift the resulting graph vertically two units upward.

7. Start with the graph of the Cube Root function and shift it horizontally four units to the left. Shrink the resulting graph toward the x-axis with a shrink factor of 0.6. Shift the new graph vertically five units downward.

8. Start with the graph of the Greatest Integer function and shift it horizontally one and one-half units to the left. Reflect the resulting graph about the x-axis. Shift the new graph vertically five units downward.

9. Rewrite the equation as $y = M(x) = -4(x+1)^3 + 3$. Start with the graph of the Cubing function and shift it horizontally one unit to the left. Stretch the resulting graph away from the x-axis with a stretch factor of 4. Reflect the new graph about the x-axis, and then shift it vertically three units upward.

10. Start with the graph of the Squaring function and shift it horizontally two units to the right. Stretch the resulting graph away from the x-axis with a stretch factor of 4. Reflect the new graph about the x-axis and then shift it vertically seven units upward.

5

Linear and Quadratic Polynomial Functions

In this chapter, we consider linear and quadratic polynomial functions and their graphs. The zeros of these polynomials will be discussed. Some of the analytic geometry associated with both linear and quadratic polynomial functions will be introduced.

5.1 LINEAR POLYNOMIAL FUNCTIONS

In section 3.4, we briefly introduced linear polynomial functions. A linear polynomial function, denoted by L, is defined by

$$y = L(x) = mx + b,$$

where m and b are real numbers and $m \neq 0$.

Linear polynomial functions have the following properties:

- Every linear polynomial function is a mapping from R onto R.

- The domain of every linear polynomial function is $(-\infty, +\infty)$.

- The range of every linear polynomial function is $(-\infty, +\infty)$.

- The graph of every linear polynomial function is an oblique line.

- Every linear polynomial function is one-to-one.

- Every linear polynomial function has an inverse as a function.

It should also be noted that the Identity function is a linear polynomial function such that $a = 1$ and $b = 0$. In fact, it is the most basic of the linear polynomial functions.

EXERCISE 5.1

Graph the function A defined by $A(x) = 2x + 3$.

SOLUTION 5.1

Since A is a linear polynomial function, its graph is an oblique line. Hence, we need two distinct points on its graph to determine the line. We determine that $A(-2) = 2(-2) + 3 = -1$ and that $A(0) = 2(0) + 3 = 3$. Hence, the line passes through the points $(-2, -1)$ and $(0, 3)$. The graph of A is given in Figure 5.1.

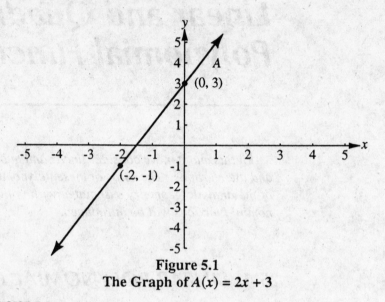

Figure 5.1
The Graph of $A(x) = 2x + 3$

EXERCISE 5.2

Graph the function B defined by $B(x) = -3x + 2$.

SOLUTION 5.2

Since B is a linear polynomial function, its graph is an oblique line. Hence, we need two distinct points on its graph to determine the line. We determine that $B(0) = -3(0) + 2 = 2$ and that $B(2) = -3(2) + 2 = -4$. Hence, the line passes through the points $(0, 2)$ and $(2, -4)$. The graph of B is given in Figure 5.2.

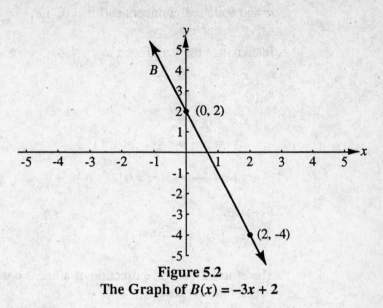

Figure 5.2
The Graph of $B(x) = -3x + 2$

EXERCISE 5.3

Determine the inverse function for the function D defined by $D(x) = 4x - 1$.

SOLUTION 5.3

D is a linear polynomial function with $a = 4$ and $b = -1$. Let $y = 4x - 1$.

Interchange the variables x and y.

$$x = 4y - 1$$

Solve for y.

$$x = 4y - 1$$

$$4y = x + 1$$

$$y = \frac{x + 1}{4}$$

Rename.

$$D^{-1}(x) = \frac{x + 1}{4} \text{ for all real values of } x.$$

EXERCISE 5.4

Determine whether the inverse function of a linear polynomial function is a linear polynomial function.

SOLUTION 5.4

Let L be a linear polynomial function defined by $L(x) = mx + b$ where

m and b are real numbers and $m \neq 0$. Let
$$y = mx + b.$$
Interchange the variables x and y.
$$x = my + b$$
Solve for y.
$$x = my + b$$
$$my = x - b$$
$$y = \frac{x - b}{m} \quad (m \neq 0)$$
Rename.
$$L^{-1}(x) = \frac{x - b}{m}$$

Therefore, the inverse function of a linear polynomial function is a linear polynomial function.

EXERCISE 5.5

Determine the inverse function of the function E defined by $E(x) = 3x - 4$. Graph both graphs on the same set of coordinate axes.

SOLUTION 5.5

Starting with $E(x) = 3x - 4$, let
$$y = 3x - 4.$$
Interchange the variables x and y.
$$x = 3y - 4$$
Solve for y.
$$x = 3y - 4$$
$$3y = x + 4$$
$$y = \frac{x + 4}{3}$$
Rename.
$$E^{-1}(x) = \frac{x + 4}{3}$$

The graphs of E and E^{-1} are given in Figure 5.3.

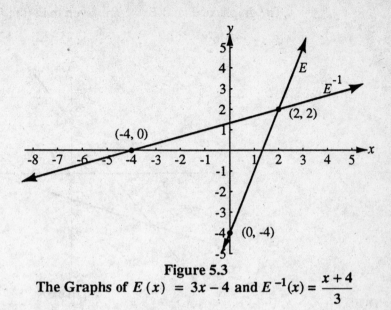

Figure 5.3
The Graphs of $E(x) = 3x - 4$ and $E^{-1}(x) = \dfrac{x+4}{3}$

EXERCISE 5.6

Determine the inverse function for the function F defined by $F(x) = \dfrac{2x-5}{4}$. Graph both functions on the same set of coordinate axes.

SOLUTION 5.6

Starting with $F(x) = \dfrac{2x-5}{4}$, let

$$y = \frac{2x-5}{4}.$$

Interchange the variables x and y.

$$x = \frac{2y-5}{4}$$

Solve for y.

$$x = \frac{2y-5}{4}$$

$$4x = 2y - 5$$

$$2y = 4x + 5$$

$$y = \frac{4x+5}{2}$$

Rename.

$$F^{-1}(x) = \frac{4x+5}{2}$$

The graphs of F and F^{-1} are given in Figure 5.4.

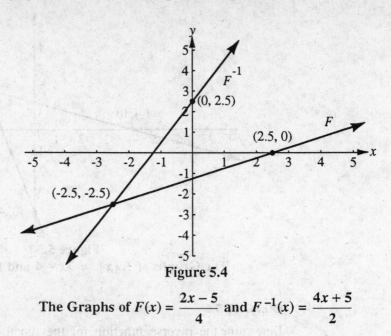

Figure 5.4

The Graphs of $F(x) = \dfrac{2x-5}{4}$ and $F^{-1}(x) = \dfrac{4x+5}{2}$

5.2 SLOPE OF A LINE

The direction of the graph of a linear function $L(x) = mx + b$ $(m \neq 0)$ is determined by the value of m. If $m > 0$, then L is an increasing function and its graph rises from left to right. If $m < 0$, then L is a decreasing function and its graph falls from left to right. The value of m determines the slope of the line.

Slope of an Oblique Line

> **Definition:**
> Let $P_1(x_1, y_1)$ and $P_2(x_2, y_2)$ be two distinct points on the line determined by the equation $y = mx + b$ $(m \neq 0)$. Then, the **slope**, m, of the line is given by
>
> $$m = \frac{y_2 - y_1}{x_2 - x_1} \qquad (x_1 \neq x_2)$$

EXERCISE 5.7

Determine the slope of the line given by the equation $y = 2x + 7$.

SOLUTION 5.7

Let $x_1 = -2$ and $x_2 = 3$. Then, $y_1 = 2\,(-2) + 7 = 3$ and $y_2 = 2\,(3) + 7 = 13$. Hence,

$$m = \frac{y_2 - y_1}{x_2 - x_1} = \frac{13 - 3}{3 - (-2)}$$

$$= \frac{10}{5}$$

$$= 2$$

EXERCISE 5.8

Determine the slope of the line given by the equation $y = -3x - 2$.

SOLUTION 5.8

Let $x_1 = -1$ and $x_2 = 4$. Then, $y_1 = -3\,(-1) - 2 = 1$ and $y_2 = -3\,(4) - 2 = -14$. Hence,

$$m = \frac{y_2 - y_1}{x_2 - x_1} = \frac{-14 - 1}{4 - (-1)}$$

$$= \frac{-15}{5}$$

$$= -3$$

EXERCISE 5.9

Show that the slope of the line given by the equation $y = ax + b$ $(a \neq 0)$ is $m = a$.

SOLUTION 5.9

Since y is defined for all real values of x, let $x_1 = -3$ and $x_2 = 1$. Then, $y_1 = a\,(-3) + b = -3a + b$ and $y_2 = a\,(1) + b = a + b$. Hence,

$$m = \frac{y_2 - y_1}{x_2 - x_1} = \frac{(a + b) - (-3a + b)}{1 - (-3)}$$

$$= \frac{a + b + 3a - b}{1 + 3}$$

$$= \frac{4a}{4}$$

$$= a.$$

Therefore, the slope, m, is equal to a.

Slope of a Horizontal Line

In section 3.3, we introduced Constant-Valued functions and noted that their graphs are horizontal lines. All points on the same horizontal line have the same ordinate values. That is, in the xy-plane, all points on the same horizontal line have the same y values. The slope of a horizontal line is 0.

EXERCISE 5.10

Determine the slope of the line, in the xy-plane, that passes through the two points $A(-3, 2)$ and $B(5, 2)$.

SOLUTION 5.10

Let $x_1 = -3$, $y_1 = 2$, $x_2 = 5$, and $y_2 = 2$. Hence,

$$m = \frac{y_2 - y_1}{x_2 - x_1} = \frac{2 - 2}{5 - (-3)}$$

$$= \frac{0}{8}$$

$$= 0.$$

In Exercise 5.10, the points A and B have the same ordinates. Hence, the line passing through the two points is a horizontal line. Its slope is 0. A horizontal line is *not* the graph of a linear function. However, it is the graph of a function — a Constant-Valued function.

Slope of a Vertical Line

A vertical line is not the graph of a function. All points on the same vertical line have the same abscissa values. That is, in the xy-plane, all points on the same vertical line have the same x values. The slope of a vertical line is not defined.

EXERCISE 5.11

Determine the slope of the line that passes through the two points $C(3, -5)$ and $D(3, 7)$.

SOLUTION 5.11

Let $x_1 = 3$, $y_1 = -5$, $x_2 = 3$, and $y_2 = 7$. Hence,

$$m = \frac{y_2 - y_1}{x_2 - x_1} = \frac{7 - (-5)}{3 - 3}$$

$$= \frac{12}{0}$$

which is not defined.

In Exercise 5.11, the points C and D lie on the same vertical line in the xy-plane since their abscissas are the same. A vertical line is not the graph of a function. The slope of a vertical line is not defined.

5.3 EQUATIONS OF LINES

There are several forms for the equation of a line. These will be considered in this section.

Point–Slope Form   Let L be the line passing through the two points $P_1(x_1, y_1)$ and $P_2(x_2, y_2)$ in the xy-plane, such that $x_1 \neq x_2$. Let $P(x, y)$ be an arbitrary point on the line L such that P is different from P_1 and P_2. We determine the slope of L as follows:

$$m = \frac{y_2 - y_1}{x_2 - x_1}.$$

Since the point $P(x, y)$ also lies on the line L, then the slope of the line through the two points $P_1(x_1, y_1)$ and $P(x, y)$ must also be equal to m. Hence,

$$m = \frac{y - y_1}{x - x_1}.$$

Since $x \neq x_1$, then $x - x_1 \neq 0$, and we can multiply both sides of the last equation by $x - x_1$, obtaining

$$y - y_1 = m(x - x_1) \tag{1}$$

Equation 1 is called the **point–slope form** of the equation of the line L since it gives the equation of the line in terms of the coordinates of the point $P_1(x_1, y_1)$, which lies on the line L, and the slope, m.

EXERCISE 5.12

Determine the equation of the line, in the xy-plane, that passes through the point $A(2, -3)$ and has a slope equal to -2.

SOLUTION 5.12

Since we know the slope of the line and a point through which it passes, we use the point–slope form of the equation of the line, with $m = -2$, $x_1 = 2$, and $y_1 = -3$. Hence,

$$y - y_1 = m(x - x_1)$$
$$y - (-3) = -2(x - 2)$$
$$y + 3 = -2x + 4$$
$$y = -2x + 1$$

The required equation of the line, then, is $y = -2x + 1$.

EXERCISE 5.13

Determine the equation of the line, in the xy-plane, that passes through the point $B(-4, -5)$ and has a slope equal to 0.3.

SOLUTION 5.13

We use the point-slope form for the equation of the line, with $m = 0.3$, $x_1 = -4$, and $y_1 = -5$. Hence,

$$y - y_1 = m(x - x_1)$$
$$y - (-5) = 0.3[x - (-4)]$$
$$y + 5 = 0.3(x + 4)$$
$$y + 5 = 0.3x + 1.2$$
$$y = 0.3x - 3.8$$

The required equation of the line, then, is $y = 0.3x - 3.8$.

EXERCISE 5.14

Determine the equation of the line, in the xy-plane, that passes through the two points $C(2, -3)$ and $D(-4, 5)$.

SOLUTION 5.14

Since we know two points through which the line passes, we can determine its slope.

$$m = \frac{5 - (-3)}{-4 - 2}$$
$$= \frac{8}{-6}$$
$$= \frac{-4}{3}$$

Now, knowing the slope of the line, we can use either the point C or the point D for the given point. We will use the point C, with $x = 2$ and $y = -3$. Using the point–slope form of the equation of the line, we have

$$y - y_1 = m(x - x_1)$$

$$y - (-3) = \frac{-4}{3}(x - 2)$$

$$3(y + 3) = -4(x - 2)$$

$$3y + 9 = -4x + 8$$

$$4x + 3y = -1$$

The required equation for the line, then, is $y = \dfrac{-4x - 1}{3}$.

Slope-intercept Form

We have noted that the graph of every linear polynomial function is an oblique line. Every oblique line in the xy-plane will cross or intersect the x-axis and also the y-axis.

> **Definition:**
>
> If a line in the xy-plane crosses, or intersects, the x-axis at the point $(a, 0)$, then a is called the **x-intercept**. If the line crosses, or intersects, the y-axis at the point $(0, b)$, then b is called the **y-intercept**.

EXERCISE 5.15

Determine both the x-intercept and the y-intercept for the line with equation $y = 4x - 5$.

SOLUTION 5.15

To determine the x-intercept, we set y equal to 0. Hence,

$$y = 4x - 5$$

$$0 = 4x - 5$$

$$4x = 5$$

$$x = \frac{5}{4}$$

The x-intercept is $\dfrac{5}{4}$. The line crosses the x-axis at the point $\left(\dfrac{5}{4}, 0\right)$.

To determine the y-intercept, we set x equal to 0. Hence,

$$y = 4x - 5$$

$$y = 4(0) - 5$$

$$y = -5$$

The y-intercept is -5. The line crosses the y-axis at the point $(0, -5)$.

EXERCISE 5.16

Determine both the x-intercept and the y-intercept for the line with equation $2x - 3y = 6$.

SOLUTION 5.16

To determine the x-intercept, we set y equal to 0. Hence,

$$2x - 3y = 6$$
$$2x - 3(0) = 6$$
$$2x - 0 = 6$$
$$2x = 6$$
$$x = 3$$

The x-intercept is 3. The line crosses the x-axis at the point $(3, 0)$.

To determine the y-intercept, we set x equal to 0. Hence,

$$2x - 3y = 6$$
$$2(0) - 3y = 6$$
$$0 - 3y = 6$$
$$-3y = 6$$
$$y = -2$$

The y-intercept is -2. The line crosses the y-axis at the point $(0, -2)$.

It should be noted that a horizontal line in the xy-plane has a y-intercept but no x-intercept, unless, of course, the line is the x-axis. Similarly, a vertical line in the xy-plane has an x-intercept but no y-intercept, unless the line is the y-axis.

Now, returning to the point–slope form of the equation of a line in the xy-plane, we observe that the point $(0, b)$ lies on the line. Hence, letting $x_1 = 0$ and $y_1 = b$, we have

$$y - y_1 = m(x - x_1)$$
$$y - b = m(x - 0)$$
$$y - b = mx$$
$$y = mx + b \qquad (2)$$

Equation 2 is called the **slope–intercept** form of the equation of a non-vertical line since it is given in terms of the slope, m, of the line and the y-intercept, b, of the line.

EXERCISE 5.17

Determine the equation of the line, in the xy-plane, that has a slope of -7 and a y-intercept of 1.

SOLUTION 5.17

Using the slope–intercept form of the equation of a line with $m = -7$ and $b = 1$, we have

$$y = mx + b$$

$$y = -7x + 1$$

The required equation is $y = -7x + 1$.

EXERCISE 5.18

Determine the equation of the line, in the xy-plane, that has a slope of 4 and passes through the point $A(0, -6)$.

SOLUTION 5.18

Since the point $A(0, -6)$ lies on the y-axis, the y-intercept is equal to -6. Using the slope–intercept form of the equation of the line with $m = 4$ and $b = -6$, we have

$$y = mx + b$$

$$y = 4x - 6.$$

The required equation is $y = 4x - 6$.

EXERCISE 5.19

Determine the slope of the line with equation $2x - 7y = 6$.

SOLUTION 5.19

Solving the given equation for y, we have

$$2x - 7y = 6$$

$$-7y = -2x + 6$$

$$y = \frac{2}{7}x - \frac{6}{7}.$$

Comparing this last equation with the slope–intercept form of the equation of a line, we have that $m = \frac{2}{7}$ and $b = \frac{-6}{7}$. Hence, the slope of the given line is $\frac{2}{7}$.

The Two-intercepts Form

Consider the line, in the xy-plane, that passes through the two points $A(a, 0)$ and $B(0, b)$ such that $a \neq 0$ and $b \neq 0$. Then, the slope of the line is determined as follows:

$$m = \frac{b - 0}{0 - a}$$

$$= \frac{-b}{a}$$

Knowing the slope of the line and a point through which it passes, we can use the point–slope form of the equation of the line as follows:

$$y - y_1 = m(x - x_1)$$

$$y - 0 = \frac{-b}{a}(x - a) \qquad \text{(Using the point } A)$$

$$a(y - 0) = -b(x - a) \qquad \text{(Since } a \neq 0)$$

$$ay = -bx + ab$$

$$bx + ay = ab$$

Now, since $a \neq 0$ and $b \neq 0$, we can divide both sides of the last equation by $ab \neq 0$, obtaining

$$\frac{bx}{ab} + \frac{ay}{ab} = \frac{ab}{ab}$$

or

$$\frac{x}{a} + \frac{y}{b} = 1 \qquad\qquad (3)$$

Equation 3 is called the **two-intercepts** form of the equation of a line since it is given in terms of the x-intercept, a, and the y-intercept, b, provided that *both a and b* are not equal to 0.

EXERCISE 5.20

Determine the equation of the line, in the xy-plane, with x-intercept = 3 and y-intercept = –4.

SOLUTION 5.20

Using the two-intercepts form for the equation of a line with $a = 3$ and $b = -4$, we have

$$\frac{x}{a} + \frac{y}{b} = 1$$

$$\frac{x}{3} + \frac{y}{-4} = 1$$

$$-4x + 3y = -12 \qquad \text{(Multiply both sides by } -12)$$
$$4x - 3y = 12 \qquad \text{(Multiply both sides by } -1)$$

EXERCISE 5.21

Determine the slope–intercept form of the equation of a line, in the xy-plane, that has x-intercept = –2 and y-intercept = 5.

SOLUTION 5.21

Using the two-intercepts form of the equation of a line with $a = -2$ and $y = 5$, we have

$$\frac{x}{a} + \frac{y}{b} = 1$$

$$\frac{x}{-2} + \frac{y}{5} = 1$$

$$5x - 2y = -10 \qquad \text{(Multiply both sides by } -10\text{)}$$

To get the equation in the slope–intercept form, we solve for y.

$$5x - 2y = -10$$

$$-2y = -5x - 10$$

$$y = \frac{5}{2}x + 5$$

The required equation is $y = \frac{5}{2}x + 5$.

Caution:

The two-intercepts form of the equation of a line can only be used if *both* intercepts are different from 0.

General Form

Equations 1, 2, and 3 can each be written in the form

$$Ax + By + C = 0 \qquad (4)$$

where A, B, and C are constants such that at least one of the values A or B is not equal to 0. Equation 4 is called the **general form** of the equation of a line in the xy-plane.

In writing the general form of the equation of a line, it is customary to write the equation with a positive coefficient for x and with integer coefficients.

EXERCISE 5.22

Determine the general form of the equation of the line, in the xy-plane, with $m = -3$ and $b = 2$.

SOLUTION 5.22

Since we are given the slope of the line and its y-intercept, we use the slope–intercept form of the equation of the line as follows:

$$y = mx + b$$

$$y = -3x + 2$$

We now rewrite the last equation in the general form.

$$y = -3x + 2$$

$$3x + y = 2$$
$$3x + y - 2 = 0$$

with $A = 3$, $B = 1$, and $C = -2$.

EXERCISE 5.23

Determine the general form of the equation of the line, in the xy-plane, passing through the points $A(2, 3)$ and $B(-3, -5)$.

SOLUTION 5.23

Since we are given two points through which the line passes, we can determine the slope of the line.

$$m = \frac{-5 - 3}{-3 - 2}$$
$$= \frac{-8}{-5}$$
$$= \frac{8}{5}$$

Knowing the slope of the line and a point through which it passes, we use the point–slope form of the equation of the line as follows:

$$y - y_1 = m(x - x_1)$$
$$y - 3 = \frac{8}{5}(x - 2) \qquad \text{(Using the point } A)$$

The general form of the equation of the line is obtained as follows:

$$y - 3 = \frac{8}{5}(x - 2)$$
$$5(y - 3) = 8(x - 2)$$
$$5y - 15 = 8x - 16$$
$$-8x + 5y - 15 = -16$$
$$-8x + 5y + 1 = 0$$
$$8x - 5y - 1 = 0$$

The required equation is $8x - 5y - 1 = 0$.

In Table 5.1, a summary of the general form of the equation of a line is given, according to whether A is nonzero or B is nonzero.

Table 5.1

Consider the equation $Ax + By + C = 0$.
- If $A \neq 0$ and $B = 0$, then the equation becomes $Ax + C = 0$, or x

$= \dfrac{-C}{A}$, for all real values of y. The equation describes a *vertical*

line in the xy-plane. The slope is not defined. The line is not the graph of a function.
- If $A = 0$ and $B \neq 0$, then the equation becomes $By + C = 0$, or

$y = \dfrac{-C}{B}$, for all real values of x. The equation describes a *hori-*

zontal line in the xy-plane. The slope is 0. The line is the graph of a Constant-Valued function.
- If $A \neq 0$ and $B \neq 0$, then the equation describes an *oblique* line in the xy-plane. It is the graph of a linear function.
  i) If A and B both have the same algebraic sign, then the slope of the line is negative.
  ii) If A and B have opposite algebraic signs, then the slope of the line is positive.

EXERCISE 5.24

Each of the following equations describes a line in the xy-plane. Write the general form of the equation of the line and determine whether the line is vertical, horizontal, or oblique. Also, determine whether the slope of the line is positive, negative, zero, or not defined.

a) $2x - y = 0$

b) $y - 5 = 0$

c) $x - y = 5$

d) $6 - 2x = 5$

e) $y = 3x - 2$

f) $2y - x + 5 = 0$

g) $3 = x + y$

h) $x = 9$

i) $7 - 4x = 3y$

j) $9 - 4y = 2x$

SOLUTION 5.24

a) The equation $2x - y = 0$ is in the general form with $A = 2$, $B = -1$, and $C = 0$. The line is oblique. Since A and B have opposite signs, the slope of the line is positive.

b) The equation $y - 5 = 0$ is in the general form with $A = 0$, $B = 1$, and $C = -5$. The line is horizontal and its slope is 0.

c) The general form of the equation $x - y = 5$ is $x - y - 5 = 0$, with $A = 1$, $B = -1$, and $C = -5$. The line is oblique. Since A and B have opposite signs, the slope of the line is positive.

d) The general form of the equation $6 - 2x = 5$ is $2x - 1 = 0$, with $A = 2$, $B = 0$, and $C = -1$. The line is vertical and its slope is not defined.

e) The general form of the equation $y = 3x - 2$ is $3x - y - 2 = 0$, with $A = 3$, $B = -1$, and $C = -2$. The line is oblique. Since A and B have opposite signs, the slope of the line is positive.

f) The general form of the equation $2y - x + 5 = 0$ is $x - 2y - 5 = 0$, with $A = 1$, $B = -2$, and $C = -5$. The line is oblique. Since A and B have opposite signs, the slope of the line is positive.

g) The general form of the equation $3 = x + y$ is $x + y - 3 = 0$, with $A = 1$, $B = 1$, and $C = -3$. The line is oblique. Since A and B have the same sign, the slope of the line is negative.

h) The general form of the equation $x = 9$ is $x - 9 = 0$ with $A = 1$, $B = 0$, and $C = -9$. The line is vertical and its slope is not defined.

i) The general form of the equation $7 - 4x = 3y$ is $4x + 3y - 7 = 0$, with $A = 4$, $B = 3$, and $C = -7$. The line is oblique. Since A and B have the same sign, the slope of the line is negative.

j) The general form of the equation $9 - 4y = 2x$ is $2x + 4y - 9 = 0$, with $A = 2$, $B = 4$, and $C = -9$. The line is oblique. Since A and B have the same sign, the slope of the line is negative.

5.4 PARALLEL AND PERPENDICULAR LINES

By an examination of the slopes of two nonvertical lines, we can determine if the lines are parallel.

Definition:

Two *nonvertical* lines, in the same plane, with slopes m_1 and m_2 are **parallel** if and only if $m_1 = m_2$. Distinct *vertical* lines, in the same plane, are also parallel.

By an examination of the slopes of two nonvertical lines, we can also determine if the lines are perpendicular.

Definition:

Two *nonvertical* lines, in the same plane, with slopes m_1 and m_2 are **perpendicular** if and only if

$$m_1 = \frac{-1}{m_2}.$$

It should be noted that a vertical line and a horizontal line, in the same plane, are also perpendicular. However, the slope of the vertical line is not defined and the slope of the horizontal line is 0.

EXERCISE 5.25

Determine whether the lines given by the following pairs of equations are parallel, perpendicular, or neither.

a) $2x - 5y + 1 = 0$ and $5x + 2y = 0$

b) $x - 2y = 7$ and $2x - 3 = 4y$

c) $x - 5 = 0$ and $2y - 8 = 0$

d) $2x - 3y = 6$ and $y - 7 = 3x$

e) $x - 3y = 1$ and $2x - 2 = 6y$

SOLUTION 5.25

a) Rewriting both equations in the slope–intercept form, we have

$$2x - 5y + 1 = 0 \qquad \text{and} \qquad 5x + 2y = 0$$

$$5y = 2x + 1 \qquad\qquad\qquad 2y = -5x$$

$$y = \frac{2}{5}x + \frac{1}{5} \qquad\qquad\qquad y = \frac{-5}{2}x$$

$$\left(m_1 = \frac{2}{5}\right) \qquad\qquad\qquad \left(m_2 = \frac{-5}{2}\right)$$

Since $m_1 = \dfrac{-1}{m_2}$, the lines are perpendicular.

b) Rewriting both equations in the slope–intercept form, we have

$$x - 2y = 7 \qquad \text{and} \qquad 2x - 3 = 4y$$

$$-2y = -x + 7 \qquad\qquad\qquad 4y = 2x - 3$$

$$y = \frac{1}{2}x - \frac{7}{2} \qquad\qquad\qquad y = \frac{1}{2}x - \frac{3}{4}$$

$$\left(m_1 = \frac{1}{2}\right) \qquad\qquad \left(m_2 = \frac{1}{2}\right)$$

Since $m_1 = m_2$, the lines are parallel.

c) The equation $x - 5 = 0$ can be rewritten as $x = 5$. The equation describes a vertical line. The equation $2y - 8 = 0$ can be rewritten as $y = 4$. The equation describes a horizontal line. Hence, the lines are perpendicular.

d) Rewriting both equations in the slope–intercept form, we have

$$2x - 3y = 6 \qquad\text{and}\qquad y - 7 = 3x$$
$$-3y = -2x + 6 \qquad\qquad y = 3x + 7$$

$$y = \frac{2}{3}x - 2$$

$$\left(m_1 = \frac{2}{3}\right) \qquad\qquad (m_2 = 3)$$

Since $m_1 \neq m_2$ and $m_1 \neq \dfrac{-1}{m_2}$, the two lines are neither parallel nor perpendicular.

e) Rewriting both equations in the slope–intercept form, we have

$$x - 3y = 1 \qquad\text{and}\qquad 2x - 2 = 6y$$
$$-3y = -x + 1 \qquad\qquad 6y = 2x - 2$$

$$y = \frac{1}{3}x - \frac{1}{3} \qquad\qquad y = \frac{1}{3}x - \frac{1}{3}$$

The two equations are identical. They describe the same line.

EXERCISE 5.26

Determine the equation of the line, in the xy-plane, that passes through the point $A(-1, 3)$ and is parallel to the line L with the equation $2y - 3x = 7$.

SOLUTION 5.26

If the required line is parallel to the line L, then its slope must be equal to the slope of L. Rewriting the equation for the line L, we have

$$2y - 3x = 7$$
$$2y = 3x + 7$$

$$y = \frac{3}{2}x + \frac{7}{2}$$

Hence, the slope of L is $\dfrac{3}{2}$. This is also the slope of the required line.

Now, knowing the slope of the required line and a point through which it passes, we use the point–slope form of the equation of the line, obtaining

$$y - y_1 = m(x - x_1)$$

$$y - 3 = \frac{3}{2}[x - (-1)]$$

$$2(y - 3) = 3(x + 1)$$

$$2y - 6 = 3x + 3$$

$$2y = 3x + 9$$

which is the equation for the required line.

EXERCISE 5.27

Determine the equation of the line, in the xy-plane, that is parallel to the line T with equation $2x - 3y + 4 = 0$ and has a y-intercept of 3.

SOLUTION 5.27

If the required line is parallel to the line T, then its slope must be equal to the slope of T. Rewriting the equation for the line T, we have

$$2x - 3y + 4 = 0$$

$$-3y = -2x - 4$$

$$y = \frac{2}{3}x + \frac{4}{3}$$

Hence, the slope of T is $\frac{2}{3}$. This is also the slope of the required line.

Now, knowing the slope of the required line and the y-intercept, we use the slope–intercept form of the equation of the line, obtaining

$$y = mx + b$$

$$y = \frac{2}{3}x + 3$$

which is the equation for the required line.

EXERCISE 5.28

Determine the equation of the line, in the xy-plane, that is perpendicular to the line R with equation $3x + 4y = 0$ and passes through the point $B(2, -1)$.

SOLUTION 5.28

If the required line is perpendicular to the line R, then its slope must be equal to the negative reciprocal of the slope of R. Rewriting the equation for the line R, we have

$$3x + 4y = 0$$

$$4y = -3x$$

$$y = \frac{-3}{4}x$$

Hence, the slope of R is $\frac{-3}{4}$. The slope of the required line, then, is $\frac{4}{3}$. Now, knowing the slope of the required line and a point through which it passes, we use the point–slope form of the equation of the line, obtaining

$$y - y_1 = m(x - x_1)$$

$$y - (-1) = \frac{4}{3}(x - 2)$$

$$3(y + 1) = 4(x - 2)$$

$$3y + 3 = 4x - 8$$

$$3y = 4x - 11$$

which is the equation for the required line.

5.5 QUADRATIC POLYNOMIAL FUNCTIONS

We will now discuss quadratic polynomial functions. A **quadratic polynomial function**, Q, is defined by

$$Q(x) = ax^2 + bx + c$$

where a, b, and c are real numbers and $a \neq 0$.

Quadratic polynomial functions have the following properties.
* Every quadratic polynomial function is a mapping from R into R.
* No quadratic polynomial function is a mapping from R onto R.
* The domain of every quadratic polynomial function is $(-\infty, +\infty)$.
* The range of every quadratic polynomial function is a *proper* subset of the set of real numbers, and varies according to the specification of the function. (Note: A proper subset of the set of real numbers is a set of real numbers that does not contain all of the real numbers.)
* The graph of every quadratic polynomial function is a parabola with a vertical axis.
* No quadratic polynomial function is one-to-one.
* No quadratic polynomial function has an inverse as a function.

The most basic quadratic polynomial function is the Squaring function, defined by $S(x) = x^2$ for all real values of x. Comparing x^2 to

$ax^2 + bx + c$, we have that $a = 1$, $b = 0$, and $c = 0$. The graph of S is given in Figure 5.5. The domain of S is $(-\infty, +\infty)$ and its range is $[0, +\infty)$. The graph is a parabola that opens upward about the y-axis. The y-axis, in this case, is the axis of the parabola. The axis of the parabola passes through the "low" point of the graph, which, in this case, is the point at the origin. This point is called the **vertex** of the parabola.

Next, we consider the quadratic polynomial functions such as $f(x) = 2x^2$, $g(x) = 4x^2$, $h(x) = \frac{1}{2}x^2$, and $j(x) = \frac{1}{3}x^2$. The functions f and g have graphs that are stretchings of the graph of S away from the x-axis, with stretch factors of 2 and 4, respectively. (See Figure 5.6.) The functions h and j have graphs that are shrinkings of the graph of S toward the x-axis, with shrink factors of $\frac{1}{2}$ and $\frac{1}{3}$, respectively. (See Figure 5.7.)

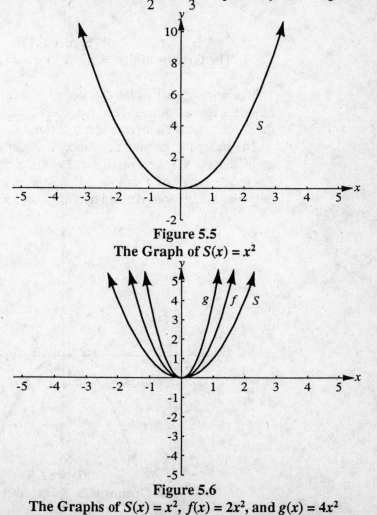

Figure 5.5
The Graph of $S(x) = x^2$

Figure 5.6
The Graphs of $S(x) = x^2$, $f(x) = 2x^2$, and $g(x) = 4x^2$

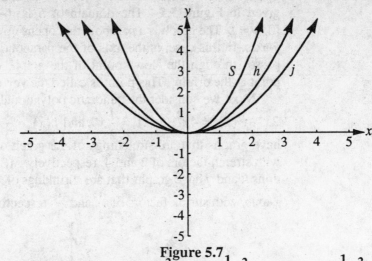

Figure 5.7
The Graphs of $S(x) = x^2$, $h(x) = \dfrac{1}{2}x^2$, and $j(x) = \dfrac{1}{3}x^2$

The range of each of the functions f, g, h, and j is $[0, +\infty)$. None of the functions is one-to-one. Their graphs fail the horizontal line test. None of the functions has an inverse as a function.

The quadratic polynomial function F defined by $F(x) = -x^2$ has a graph that is the reflection of the graph of S about the x-axis. (See Figure 5.8.) The range of F is $(-\infty, 0]$. Similarly, the range of the quadratic polynomial functions defined by $H(x) = ax^2$, where $a < 0$, is $(-\infty, 0]$.

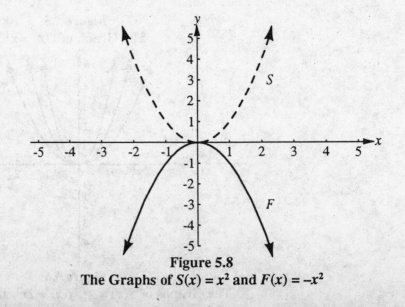

Figure 5.8
The Graphs of $S(x) = x^2$ and $F(x) = -x^2$

In chapter 4, we determined that the functions such as those defined by $y = (x-1)^2$, $y = (x+3)^2$, and $y = (x-6)^2$ all have graphs that are horizontal shifts of the graph of $S(x) = x^2$. Hence, the range of each of these functions is $[0, +\infty)$. The functions such as those defined by $y = -(x+2)^2$, $y = -(x-5)^2$, and $y = -(x+8)^2$ all have graphs that are horizontal shifts of the graph of S, followed by a reflection about the x-axis. Hence, the range of each of these functions is $(-\infty, 0]$.

The graphs of all these quadratic polynomial functions are parabolas that open either upward or downward about the vertical axis of the parabola.

Every equation of the form $y = ax^2 + bx + c$ $(a \neq 0)$ can be rewritten in the form $y = a(x-h)^2 + k$ $(a \neq 0)$. From chapter 4, we recognize that the graph of such an equation is a variation of the graph of the Squaring function. The graph may be obtained as follows:

Step 1:

Start with the graph of $S(x) = x^2$.

Step 2:

Shift the graph of S horizontally $|h|$ units, to the right if $h > 0$ or to the left if $h < 0$.

Step 3:

If $|a| > 1$, stretch the resulting graph away from the x-axis, with a stretch factor of $|a|$. If $0 < |a| < 1$, shrink the resulting graph toward the x-axis, with a shrink factor of $|a|$.

Step 4:

If $a < 0$, reflect the resulting graph about the x-axis.

Step 5:

Shift the new graph vertically $|k|$ units, upward if $k > 0$ or downward if $k < 0$.

Completing the Square

Before continuing with our discussion, we introduce the idea of completing a square on a quadratic polynomial. For instance, the expression $x^2 + 2x + 1$ is considered to be a perfect square since it can be rewritten as the square of $(x+1)$. That is, $x^2 + 2x + 1 = (x+1)^2$. The expression $x^2 - 6x + 9$ is also a perfect square since $x^2 - 6x + 9 = (x-3)^2$. The expression $x^2 + 4x$ is not a perfect square. However, by adding 4 to it, the expression becomes a perfect square, with $x^2 + 4x + 4 = (x+2)^2$.

Procedure for Completing the Square:

To complete the square on the expression $x^2 + px$:

Step 1:

Take one-half of p, the coefficient of x, obtaining $\frac{1}{2}p$.

Step 2:

Square $\frac{1}{2}p$, obtaining $\left(\frac{1}{2}p\right)^2$.

Step 3:

Add $\left(\frac{1}{2}p\right)^2$ to $x^2 + px$, obtaining $x^2 + px + \left(\frac{1}{2}p\right)^2$.

Step 4:

The result is a perfect square, since $x^2 + px + \left(\frac{1}{2}p\right)^2 = \left(x + \frac{p}{2}\right)^2$.

EXERCISE 5.29

Complete the square on each of the following expressions.

a) $x^2 + 8x$

b) $x^2 + 14x$

c) $x^2 + 17x$

SOLUTION 5.29

a) To complete the square on $x^2 + 8x$, take one-half of 8 (the coefficient of x), square it, and add the result to the expression.

$$x^2 + 8x \rightarrow x^2 + 8x + \left(\frac{8}{2}\right)^2$$

$$= x^2 + 8x + (4)^2$$

$$= (x + 4)^2$$

b) To complete the square on $x^2 + 14x$, take one-half of 14 (the coefficient of x), square it, and add the result to the expression.

$$x^2 + 14x \rightarrow x^2 + 14x + \left(\frac{14}{2}\right)^2$$

$$= x^2 + 14x + (7)^2$$

$$= (x + 7)^2$$

c) To complete the square on $x^2 + 17x$, take one-half of 17 (the coefficient of x), square it, and add the result to the expression.

$$x^2 + 17x \rightarrow x^2 + 17x + \left(\frac{17}{2}\right)^2$$

$$= \left(x + \frac{17}{2}\right)^2$$

EXERCISE 5.30

Complete the square on the expression $2x^2 + 8x$.

SOLUTION 5.30

The expression $2x^2 + 8x$ is not in the proper form for completing the square since the coefficent of x is not equal to 1. However, we can rewrite $2x^2 + 8x$ as

$$2\left(x^2 + 4x\right)$$

and complete the square within the parentheses.

Step 1: Take one-half of 4, the coefficient of x, obtaining 2.

Step 2: Square 2, obtaining 4.

Step 3: Add 4 to $x^2 + 4x$, within the parentheses, obtaining $x^2 + 4x + 4$.

We now have

$$2x^2 + 8x = 2\left(x^2 + 4x\right) \rightarrow 2\left(x^2 + 4x + 4\right)$$

$$= 2x^2 + 8x + 8$$

EXERCISE 5.31

Graph the function f defined by $f(x) = x^2 + 4x + 1$.

SOLUTION 5.31

We will rewrite $f(x)$ in the form $f(x) = a(x - h)^2 + k$. This can be done by completing the square on $x^2 + 4x + 1$ or, more correctly, on $x^2 + 4x$. We have

$$x^2 + 4x + 1 = \left(x^2 + 4x\right) + 1$$

$$= \left(x^2 + 4x + \left(\frac{4}{2}\right)^2\right) + 1 - \left(\frac{4}{2}\right)^2$$

(Notice that $\left(\frac{4}{2}\right)^2$ was *both* added and subtracted in the above expression. That is the equivalent of adding 0 to the given expression, thus not changing the value of the expression.) Continuing, we have

$$= (x + 2)^2 + 1 - 2^2$$

$$= (x + 2)^2 + (1 - 4)$$

$$= (x + 2)^2 - 3$$

Hence, $f(x) = (x + 2)^2 - 3$. To graph f, we start with the graph of the Squaring function, shift it horizontally two units to the left, and vertically shift the resulting graph three units downward. (See Figure 5.9.) The domain of f is $(-\infty, +\infty)$ and its range is $[-3, +\infty)$. The vertex of the parabola (the graph of f) is at the point $V(-2, -3)$. The axis of the parabola is the line with equation $x = -2$.

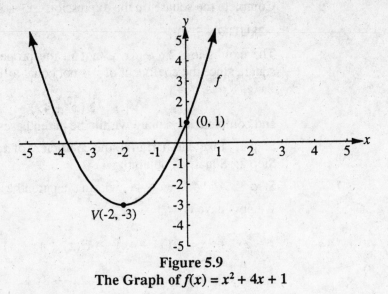

Figure 5.9
The Graph of $f(x) = x^2 + 4x + 1$

EXERCISE 5.32

The graph of the function g defined by $g(x) = ax^2 + bx + c$ $(a \neq 0)$ is a parabola with a vertical axis. Determine:
a) the coordinates of the vertex, V, of the parabola
b) the axis of the parabola

SOLUTION 5.32

Rewriting $g(x)$, we have

$$g(x) = ax^2 + bx + c \quad (a \neq 0)$$

$$= a\left(x^2 + \frac{b}{a}x\right) + c$$

$$= a\left(x^2 + \frac{b}{a}x + \left(\frac{b}{2a}\right)^2\right) + \left(c - a\left(\frac{b}{2a}\right)^2\right)$$

$$= a\left(x + \left(\frac{b}{2a}\right)\right)^2 + \left(c - \frac{ab^2}{4a^2}\right)$$

$$= a\left(x + \left(\frac{b}{2a}\right)\right)^2 + \left(c - \frac{b^2}{4a}\right)$$

$$= a\left(x + \left(\frac{b}{2a}\right)\right)^2 + \left(\frac{4ac - b^2}{4a}\right)$$

If we let $h = \dfrac{-b}{2a}$ and let $k = \dfrac{4ac - b^2}{4a}$, then this last equation can be rewritten in the form

$$g(x) = a(x - h)^2 + k.$$

The graph of g is the graph of $y = ax^2$, shifted horizontally $|h|$ units (to the right if $h > 0$, or to the left if $h < 0$). The resulting graph is then shifted vertically $|k|$ (upward if $k > 0$, or downward if $k < 0$). The parabola will open upward if $a > 0$ or downward if $a < 0$.

a) The vertex of the parabola is at the point $V(h,k)$, or $V\left(\dfrac{-b}{2a}, \dfrac{4ac - b^2}{4a}\right)$.

b) The axis of the parabola is the line $x = \dfrac{-b}{2a}$.

The Graph of a Quadratic Polynomial Function

The coordinates of the vertex of a parabola which is the graph of a quadratic polynomial function can more easily be obtained by remembering that its x-coordinate is $\dfrac{-b}{2a}$. Then, the y-coordinate can be determined by computing $g\left(\dfrac{-b}{2a}\right)$. The results of Exercise 5.32 are summarized below.

The Graph of a Quadratic Polynomial Function:
Consider the quadratic polynomial function g defined by
$$g(x) = ax^2 + bx + c \quad (a \neq 0).$$

- The graph of g is a parabola with a vertical axis.

- The vertex of the parabola is at the point $V\left(\dfrac{-b}{2a}, g\left(\dfrac{-b}{2a}\right)\right)$.

- The axis of the parabola is the vertical line with equation $x = \dfrac{-b}{2a}$.

- The parabola opens upward if $a > 0$. It opens downward if $a < 0$.

EXERCISE 5.33

Consider the function F defined by $F(x) = 3x^2 - 6x + 7$.
a) Determine the coordinates of the vertex of the graph of F.
b) Determine the equation of the axis of the parabola.
c) Determine whether the parabola opens upward or downward.

SOLUTION 5.33

$F(x)$ is of the form $ax^2 + bx + c$, with $a = 3$, $b = -6$, and $c = 7$.

a) The x-coordinate of the vertex of the parabola is $\dfrac{-b}{2a} = \dfrac{-(-6)}{2(3)} = \dfrac{6}{6} = 1$.

The y-coordinate of the vertex of the parabola is $F\left(\dfrac{-b}{2a}\right) = F(1) = 3(1)^2 - 6(1) + 7 = 3 - 6 + 7 = 4$. Hence, the vertex is at $V(1, 4)$.

b) The axis of the parabola has equation $x = \dfrac{-b}{2a}$, or $x = 1$.

c) Since $a = 3 > 0$, the parabola opens upward. (See Figure 5.10.)

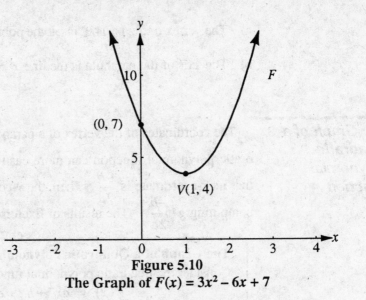

Figure 5.10
The Graph of $F(x) = 3x^2 - 6x + 7$

EXERCISE 5.34

Determine the range of the function H defined by $H(x) = -2x^2 + 8x - 3$.

SOLUTION 5.34

The graph of H is a parabola with a vertical axis. The x-coordinate of the vertex of the parabola is $\dfrac{-b}{2a} = \dfrac{-8}{2(-2)} = \dfrac{-8}{-4} = 2$. The y-coordinate of the vertex is $H(2) = -2(2)^2 + 8(2) - 3 = -8 + 16 - 3 = 5$. The vertex is at the point $V(2, 5)$. Since $a = -2 < 0$, the parabola opens downward. The range of H is $(-\infty, 5]$. That is, the range of the function is the set of all real numbers that are less than or equal to the y-coordinate of the vertex of the parabola. (See Figure 5.11.)

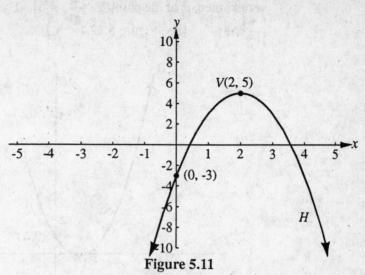

Figure 5.11
The Graph of $H(x) = -2x^2 + 8x - 3$

In general, the range of a quadratic polynomial function can be obtained as follows.

Step 1:

Determine whether the parabola (the graph of the function) opens upward or downward. This is done by examining the sign of the coefficent of the x^2.

Step 2:

Determine the coordinates of the vertex of its graph.

Step 3:

If the parabola opens upward, the range will be the set of all real numbers that are greater than or equal to the y-coordinate of the vertex.

Step 4:

If the parabola opens downward, the range will be the set of all real numbers that are less than or equal to the y-coordinate of the vertex.

EXERCISE 5.35

Determine the range of the function A defined by $y = A(x) = 2x^2 + 3x + 1$.

SOLUTION 5.35

The function A is a quadratic polynomial function. Its graph is a parabola with a vertical axis. The coefficient of x^2 is 2, which is positive. Hence, the parabola opens upward. The x-coordinate of the vertex of the parabola is given by $\dfrac{-b}{2a} = \dfrac{-3}{2(2)} = \dfrac{-3}{4}$. The y-coordinate of the vertex is

given by $A\left(\dfrac{-3}{4}\right) = 2\left(\dfrac{-3}{4}\right)^2 + 3\left(\dfrac{-3}{4}\right) + 1 = \dfrac{9}{8} - \dfrac{9}{4} + 1 = \dfrac{-1}{8}$. The vertex, then, is at the point $V\left(\dfrac{-3}{4}, \dfrac{-1}{8}\right)$. The range of the function A is $\left[\dfrac{-1}{8}, +\infty\right)$. (See Figure 5.12.)

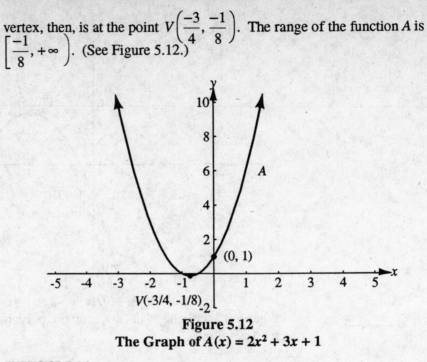

Figure 5.12
The Graph of $A(x) = 2x^2 + 3x + 1$

EXERCISE 5.36

Determine the range of the function B defined by $y = B(x) = -3x^2 + 2x - 7$.

SOLUTION 5.36

The function B is a quadratic polynomial function. Its graph is a parabola with a vertical axis. The coefficient of x^2 is -3, which is negative. Hence, the parabola opens downward. The x-coordinate of the vertex of

the parabola is given by $\dfrac{-b}{2a} = \dfrac{-2}{2(-3)} = \dfrac{1}{3}$. The y-coordinate of the

vertex is given by $B\left(\dfrac{1}{3}\right) = -3\left(\dfrac{1}{3}\right)^2 + 2\left(\dfrac{1}{3}\right) - 7 = \dfrac{-1}{3} + \dfrac{2}{3} - 7 = \dfrac{-20}{3}$.

The vertex, then, is at the point $V\left(\dfrac{1}{3}, \dfrac{-20}{3}\right)$. The range of the function B is $\left(-\infty, \dfrac{-20}{3}\right]$. (See Figure 5.13.)

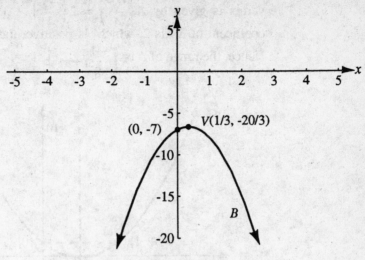

Figure 5.13
The Graph of $B(x) = -3x^2 + 2x - 7$

EXERCISE 5.37

(Alternate Approach to Exercise 5.35) Determine the range of the function A defined by $y = A(x) = 2x^2 + 3x + 1$.

SOLUTION 5.37

The function A is a quadratic polynomial function. Its graph is a parabola with a vertical axis. Since the domain of A is the set $(-\infty, +\infty)$, then $x = 0$ is in the domain. For $x = 0$, we determine that $y = A(0) = 2(0)^2 + 3(0) + 1 = 1$. Hence, 1 is in the range of A. Let $y = 1$. Then,

$$1 = 2x^2 + 3x + 1$$

or

$$2x^2 + 3x = 0.$$

Factoring, we have

$$x(2x + 3) = 0$$

from which we determine that $x = 0$ or $x = \dfrac{-3}{2}$. Hence, the points $P(0, 1)$ and $Q\left(\dfrac{-3}{2}, 1\right)$ lie on the graph of A. (See Figure 5.14.) The midpoint of the horizontal line segment connecting these two points, then, has coordinates $\left(\dfrac{-3}{4}, 1\right)$. The axis of the parabola is perpendicular to the line segment and passes through its midpoint. Hence, every point on the axis will have an x-coordinate of $\dfrac{-3}{4}$. In particular, the vertex of the parabola will have an x-coordinate of $\dfrac{-3}{4}$. The y-coordinate of the vertex is given by $A\left(\dfrac{-3}{4}\right) = 2\left(\dfrac{-3}{4}\right)^2 + 3\left(\dfrac{-3}{4}\right) + 1 = \dfrac{-1}{8}$. Since the coefficent of x^2 is 2, which is positive, the parabola will open upward. Hence, the range of A is $\left[\dfrac{-1}{8}, +\infty\right)$.

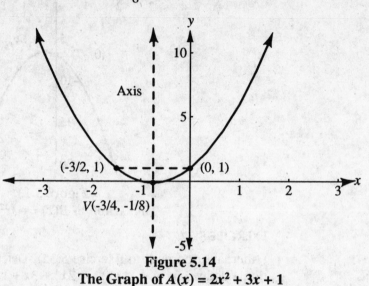

Figure 5.14
The Graph of $A(x) = 2x^2 + 3x + 1$

5.6 ZEROS OF POLYNOMIALS

We observed that the domain of every linear polynomial function and every quadratic polynomial function is the set of all real numbers. Of particular interest will be those values in the domain of the functions that are paired with 0 in their range.

Zeros of Linear Polynomials

When graphing functions, it is helpful to determine all points, if any, where the graph intersects the x-axis. The x values of such points, as noted earlier, are called the x-intercepts. If $y = f(x)$, then the x-intercepts are determined by setting $y = 0$ and solving the resulting equation $f(x) = 0$. Any solutions of the equation $f(x) = 0$ are called zeros of $f(x)$.

> **Defintion:**
> Let $f(x)$ be a linear polynomial. A **zero** of $f(x)$ is any solution of the equation $f(x) = 0$.

EXERCISE 5.38

Determine all zeros of the polynomial $ax + b$ such that $a \neq 0$.

SOLUTION 5.38

A zero of $ax + b$ $(a \neq 0)$ is any solution of the equation $ax + b = 0$. Hence,

$$ax + b = 0$$

$$ax = -b$$

$$x = \frac{-b}{a}$$

Since $a \neq 0$, $\dfrac{-b}{a}$ is a unique real number.

In Exercise 5.38, we note that *every* linear polynomial, with real coefficients, has exactly one real zero. The following relationships should be carefully noted.

Let f be a function defined by $y = f(x) = ax + b$ $(a \neq 0)$. Then:

- If x_0 is a zero of $f(x)$, then x_0 is in the domain of f and is paired with 0 in the range of f.
- x_0 is a solution of the equation $f(x) = 0$.
- x_0 is the x-coordinate of a point on the graph of f where the graph intersects the x-axis.

EXERCISE 5.39

Determine all zeros of the following linear polynomials.

a) $3x - 7$

b) $5 - 4x$

c) $8x + 11$

SOLUTION 5.39

a) Set $3x - 7 = 0$. We have

$$3x - 7 = 0$$

$$3x = 7$$

$$x = \frac{7}{3}$$

Hence, $\dfrac{7}{3}$ is a (real) zero of $3x - 7$.

b) Set $5 - 4x = 0$. We have

$$5 - 4x = 0$$

$$-4x = -5$$

$$x = \frac{5}{4}$$

Hence, $\frac{5}{4}$ is a (real) zero of $5 - 4x$.

c)   Set $8x + 11 = 0$. We have

$$8x + 11 = 0$$

$$8x = -11$$

$$x = \frac{-11}{8}$$

Hence, $\frac{-11}{8}$ is a (real) zero of $8x + 11$.

Zeros of Quadratic Polynomials

Every linear polynomial with real coefficients has exactly one real zero. As we will show, a quadratic polynomial with real coefficients may or may not have a real zero. As was noted in section 5.5, the range of a quadratic polynomial function may or may not contain 0. If 0 is not in the range of the function, then there are no real zeros for the quadratic polynomial. We will consider quadratic polynomials with real coefficients only. The following observations are now noted.

The Graph of a Quadratic Polynomial Function:

Let f be a quadratic polynomial function defined by
$f(x) = ax^2 + x + c$ $(a \neq 0)$. Then:

- Since the range of f is a proper subset of the set of real numbers, then $f(x)$ may have two, one, or no real zeros.
- If x_0 is a real zero of $f(x)$, then x_0 is in the domain of f and is paired with 0 in the range of f.
- x_0 is a solution of the quadratic equation $f(x) = 0$.
- x_0 is the x-coordinate of a point on the graph of f where the graph intersects the x-axis.
- If $f(x)$ has two real zeros, then the graph of f will cross the x-axis exactly twice.
- If $f(x)$ has exactly one real zero, then the graph of f will be tangent to the x-axis.
- If $f(x)$ has no real zeros, then the graph of f will be completely above or completely below the x-axis.

EXERCISE 5.40

Determine all real zeros, if any, of $f(x) = x^2 + 5x$.

SOLUTION 5.40

The expression $x^2 + 5x$ is factorable as $x(x+5)$. To determine the real zeros of $f(x)$, we set $f(x) = 0$. We have

$$x^2 + 5x = 0$$
$$x(x+5) = 0$$

from which we determine that

$$x = 0 \quad \text{or} \quad x + 5 = 0$$
$$x = -5$$

Therefore, $f(x)$ has two real zeros, which are -5 and 0.

EXERCISE 5.41

Determine all real zeros, if any, of $g(x) = x^2 + 5x + 6$.

SOLUTION 5.41

The expression $x^2 + 5x + 6$ is factorable as $(x+2)(x+3)$. We set $g(x) = 0$, obtaining

$$x^2 + 5x + 6 = 0$$
$$(x+2)(x+3) = 0$$
$$x + 2 = 0 \quad \text{or} \quad x + 3 = 0$$
$$x = -2 \qquad\qquad x = -3$$

Therefore, $g(x)$ has two real zeros, which are -3 and -2.

EXERCISE 5.42

Determine all real zeros, if any, of the polynomial $h(x) = 2x^2 + 3x - 2$.

SOLUTION 5.42

Set $h(x) = 0$. We have
$$2x^2 + 3x - 2 = 0$$

The left-hand side of the equation is not factorable. We will solve the equation by the method of completing the square. We have
$$2x^2 + 3x - 2 = 0$$
$$2x^2 + 3x = 2$$
$$2\left(x^2 + \frac{3}{2}x\right) = 2$$

$$\left(x^2 + \frac{3}{2}x\right) = 1$$

$$\left(x^2 + \frac{3}{2}x + \left(\frac{3}{4}\right)^2\right) = 1 + \left(\frac{3}{4}\right)^2$$

$$\left(x + \frac{3}{4}\right)^2 = 1 + \frac{9}{16}$$

$$\left(x + \frac{3}{4}\right)^2 = \frac{25}{16}$$

$$x + \frac{3}{4} = \pm\frac{5}{4}$$

$$x = \frac{-3}{4} \pm \frac{5}{4}$$

Hence,

$$x = \frac{-3}{4} + \frac{5}{4} \qquad \text{or} \qquad x = \frac{-3}{4} - \frac{5}{4}$$

$$x = \frac{2}{4} \qquad\qquad\qquad x = \frac{-8}{4}$$

$$x = \frac{1}{2} \qquad\qquad\qquad x = -2$$

Therefore, the real zeros of $h(x)$ are -2 and $\frac{1}{2}$.

EXERCISE 5.43

Determine all real zeros, if any, of the polynomial $p(x) = x^2 + 2x - 4$.

SOLUTION 5.43

Set $p(x) = 0$. We have

$$x^2 + 2x - 4 = 0$$

$$x^2 + 2x = 4$$

$$x^2 + 2x + 1 = 4 + 1 \qquad\qquad \text{(Add 1 to both sides to}$$

(Add 1 to both sides to complete the square on the left-hand side.)

$$(x + 1)^2 = 5$$

$$x + 1 = \pm\sqrt{5}$$

$$x = -1 \pm \sqrt{5}$$

Therefore, the real zeros of $p(x)$ are $-1 - \sqrt{5}$ and $-1 + \sqrt{5}$.

EXERCISE 5.44

Determine all real zeros of the quadratic polynomial $q(x) = x^2 - 10x + 25$.

SOLUTION 5.44

Set $q(x) = 0$. We have

$$x^2 - 10x + 25 = 0$$
$$(x - 5)(x - 5) = 0$$
$$(x - 5)^2 = 0$$
$$x - 5 = 0$$
$$x = 5$$

Therefore, 5 is the only real zero of $q(x)$.

EXERCISE 5.45

Determine all real zeros of the quadratic polynomial $r(x) = 4x^2 - 5x + 2 = 0$.

SOLUTION 5.45

Set $r(x) = 0$ and solve the equation using the Quadratic Formula with $a = 4$, $b = -5$, and $c = 2$. We have

$$x = \frac{-b \pm \sqrt{b^2 - 4ac}}{2a}$$

$$= \frac{-(-5) \pm \sqrt{(-5)^2 - 4(4)(2)}}{2(4)}$$

$$= \frac{5 \pm \sqrt{25 - 32}}{8}$$

$$= \frac{5 \pm \sqrt{-7}}{8}$$

Since $\sqrt{-7}$ is not a real number, $r(x)$ has no real zeros.

EXERCISE 5.46

Determine all real zeros, if any, of the quadratic polynomial $t(x) = 2x^2 - 3x + 7$.

SOLUTION 5.46

Set $t(x) = 0$ and solve the equation using the Quadratic Formula, with $a = 2$, $b = -3$, and $c = 7$.

$$x = \frac{-b \pm \sqrt{b^2 - 4ac}}{2a}$$

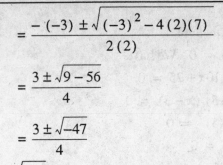

$$= \frac{-(-3) \pm \sqrt{(-3)^2 - 4(2)(7)}}{2(2)}$$

$$= \frac{3 \pm \sqrt{9 - 56}}{4}$$

$$= \frac{3 \pm \sqrt{-47}}{4}$$

Since $\sqrt{-47}$ is not a real number, $t(x)$ has no real zeros.

In this chapter, we discussed linear and quadratic polynomial functions. The characteristics of these functions and their graphs were examined. The real zeros of linear and quadratic polynomials were also discussed.

SUPPLEMENTAL EXERCISES

1. If f is a linear polynomial function, what is its domain? Its range?
2. What is the graph of a linear polynomial function?
3. What is the inverse function of the function g defined by $g(x) = 5x + 1$?
4. What is the inverse function of the function h defined by $h(x) = 4 - 3x$?
5. Determine the slope of the line given by the equation $2x - y + 5 = 0$.
6. What is the slope of: a) a horizontal line?   b) a vertical line?
7. Determine the equation of the line, in the xy-plane, that passes through the two points $A(-2, 3)$ and $B(0, 5)$.
8. Determine the x-intercept for the line with equation $3x - 2y = 9$.
9. Determine the equation of the line, in the xy-plane, that has an x-intercept $= -2$ and a y-intercept $= 4$.
10. Write the general form of the equation of the line, in the xy-plane, that has a slope of -4 and a y-intercept $= 3$.
11. Determine the equation of the line, in the xy-plane, that passes through the point $C(-1, 3)$ and is perpendicular to the line with equation $2y - 3x = 8$.
12. What is the domain of a quadratic polynomial function? Its range?
13. What is the inverse function of the function F defined by $F(x) = 3 - x^2$?
14. What is the graph of a quadratic polynomial function?
15. What are the coordinates of the vertex of the parabola that is the graph of $H(x) = 3x^2 - 4x + 7$?

16. Complete the square on $x^2 - 24x$.
17. Determine the range of the function $T(x) = 2x^2 - 8x + 3$.
18. Determine the range of the function $V(x) = -2x^2 + 4x + 2$.
19. Determine all of the real zeros, if any, of the quadratic polynomial $2x^2 + 4x - 6$.
20. Determine all of the real zeros, if any, of the quadratic polynomial $x^2 - 3x + 5$.

ANSWERS TO SUPPLEMENTAL EXERCISES

1. The set of real numbers; the set of real numbers
2. An oblique line
3. $g^{-1}(x) = \dfrac{x-1}{5}$
4. $h(x) = \dfrac{4-x}{3}$
5. $m = 2$
6. a) 0

   b) Not defined
7. $y = x + 5$
8. x-intercept $= 3$
9. $2x - y = -4$
10. $4x + y - 3 = 0$
11. $2x + 3y = 7$
12. The set of all real numbers; a proper subset of the real numbers
13. F is a quadratic polynomial function and, hence, does not have an inverse as a function.
14. A parabola with a vertical axis
15. $V\left(\dfrac{2}{3}, \dfrac{17}{3}\right)$
16. $x^2 - 24x + 144$
17. $[-5, +\infty)$
18. $(-\infty, 4]$
19. -3 and 1
20. There are none.

6

Higher-Degree Polynomial and Rational Functions

*I*n this chapter, we consider higher-degree polynomial functions and the zeros of these polynomials. Determining the zeros of higher-degree polynomials, together with an examination of their characteristics, will enable us to get the general shapes of the graphs of the respective polynomial functions. However, you will have to wait until a calculus course to get the exact shape of the graphs of these functions.

We will also introduce rational functions as the quotients of two polynomial functions.

6.1 POWER FUNCTIONS

Functions of the form $F(x) = x^n$ where n is a *positive* integer greater than or equal to 2 are called **Power** functions. In general, a Power function has a domain that is the set R. As we will see, the range of a Power function depends upon whether n is odd or even.

If $n = 2$, then $F(x) = x^2$, which is the Squaring function. Its graph is a parabola with a vertical axis. The vertex of the parabola is at the point $V(0, 0)$. The range of the function is $[0, +\infty)$. (See Figure 6.1.) The function is not one-to-one and does not have an inverse as a function.

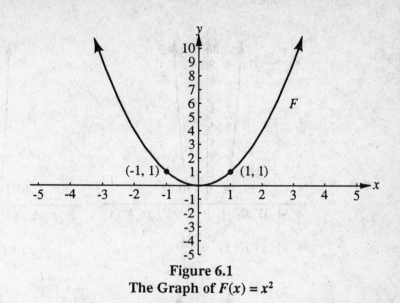

Figure 6.1
The Graph of $F(x) = x^2$

If $n = 3$, then $F(x) = x^3$, which is the Cubing function. The range of the function is $(-\infty, +\infty)$. (See Figure 6.2.) The function is one-to-one and does have an inverse as a function.

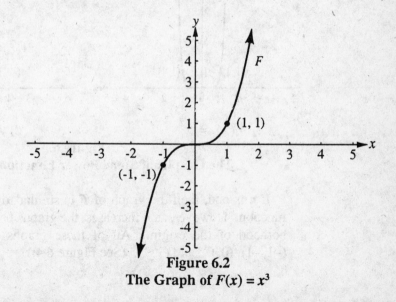

Figure 6.2
The Graph of $F(x) = x^3$

If n is *even*, the graph of F resembles the graph of the Squaring function. However, as the value of n increases, the graph of the function becomes flatter in the neighborhood of the origin. All of these functions have graphs that pass through the points $(-1, 1)$, $(0, 0)$, and $(1, 1)$. (See Figure 6.3.)

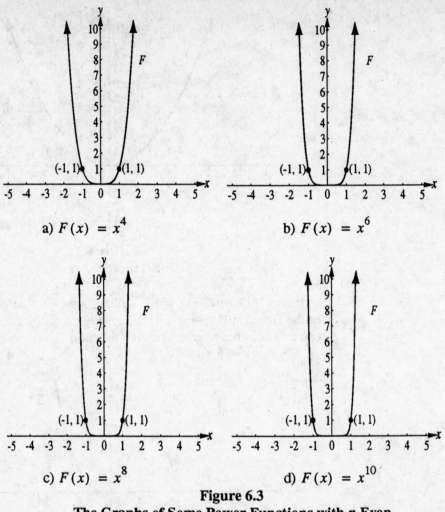

a) $F(x) = x^4$

b) $F(x) = x^6$

c) $F(x) = x^8$

d) $F(x) = x^{10}$

Figure 6.3
The Graphs of Some Power Functions with *n* Even

If *n* is odd, then the graph of *F* is similar to the graph of the Cubing function. However, as *n* increases, the graphs become flatter in the neighborhood of the origin. All of these graphs pass through the points $(-1, -1)$, $(0, 0)$, and $(1, 1)$. (See Figure 6.4.)

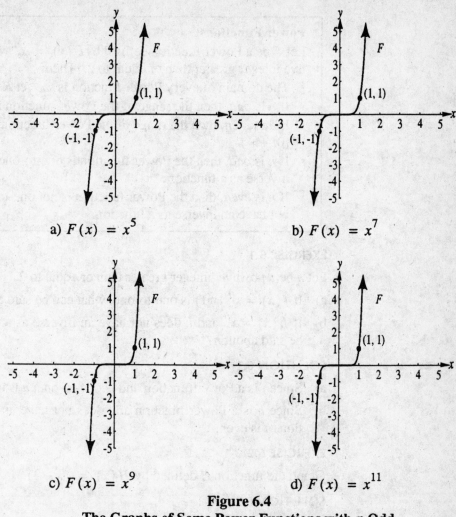

a) $F(x) = x^5$

b) $F(x) = x^7$

c) $F(x) = x^9$

d) $F(x) = x^{11}$

Figure 6.4
The Graphs of Some Power Functions with n Odd

Power Functions:

Let F be a Power function defined by $F(x) = x^n$ where n is a positive integer greater than or equal to 2. Then:

- The domain of every Power function is the set R.
- If n is *odd*, then the range of the Power function is the set R.
- If n is *even*, then the range of the Power function is the set $[0, +\infty)$.
- If n is *odd*, then the Power function is one-to-one and has an inverse as a function.
- If n is *even*, then the Power function is not one-to-one and does not have an inverse as a function.

EXERCISE 6.1

Let n be a *positive* integer greater than or equal to 2.

a) If $f(x) = x^n$ and f is one-to-one, what can be said about n?

b) If $h(x) = x^n$ and h does not have an inverse as a function, what can be said about n?

SOLUTION 6.1

a) Since f is a Power function and is one-to-one, n is odd.

b) Since h is a Power function and does not have an inverse as a function, n is even.

EXERCISE 6.2

Graph the function H defined by $H(x) = (x-3)^4$.

SOLUTION 6.2

The graph of H can be obtained by starting with the graph of the Power function $f(x) = x^4$ and shifting it horizontally 3 units to the right. The range of H is $[0, +\infty)$. (See Figure 6.5.)

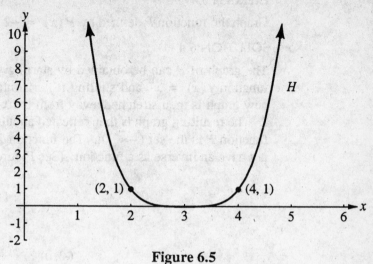

Figure 6.5
The Graph of $H(x) = (x - 3)^4$

EXERCISE 6.3

Graph the function T defined by $T(x) = x^5 + 2$.

SOLUTION 6.3

The graph of T can be obtained by starting with the graph of the Power function $f(x) = x^5$ and shifting it vertically 2 units upward. The range of $T = (-\infty, +\infty)$. The function T is one-to-one and has an inverse as a function. (See Figure 6.6.)

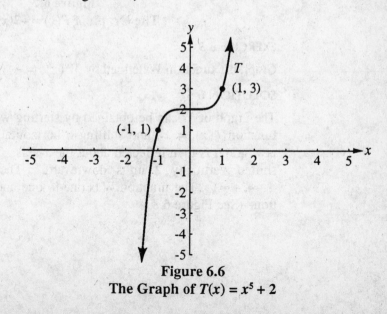

Figure 6.6
The Graph of $T(x) = x^5 + 2$

EXERCISE 6.4

Graph the function P defined by $P(x) = -2(x+3)^6$.

SOLUTION 6.4

The graph of P can be obtained by starting with the graph of the Power function $f(x) = x^6$ and shifting it horizontally 3 units to the left. The new graph is then stretched away from the x-axis with a stretch factor of 2. The resulting graph is then reflected about the x-axis. The range of the function P is the set $(-\infty, 0]$. The function P is not one-to-one and does not have an inverse as a function. (See Figure 6.7.)

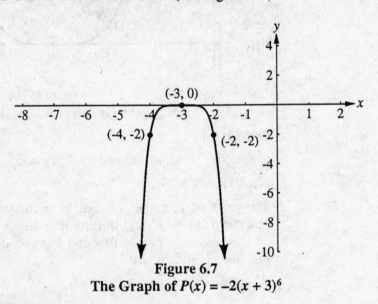

Figure 6.7
The Graph of $P(x) = -2(x+3)^6$

EXERCISE 6.5

Graph the function W defined by $W(x) = -(x+4)^5 - 2$.

SOLUTION 6.5

The graph of W can be obtained by starting with the graph of the Power function $f(x) = x^5$ and shifting it horizontally 4 units to the left. The new graph is then reflected about the x-axis. The resulting graph is then shifted vertically 2 units downward. The range of W is the set $(-\infty, +\infty)$. The function W is one-to-one and has an inverse as a function. (See Figure 6.8.)

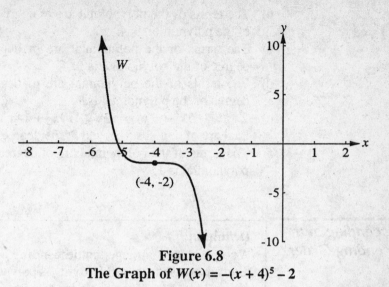

Figure 6.8
The Graph of $W(x) = -(x+4)^5 - 2$

6.2 HIGHER-DEGREE POLYNOMIALS

In general, an expression of the form

$$a_n x^n + a_{n-1} x^{n-1} + a_{n-2} x^{n-2} + \ldots + a_1 x + a_0$$

where n is a *non-negative* integer and $a_n, a_{n-1}, a_{n-2}, \ldots, a_1$, and a_0 are real numbers, with $a_n \neq 0$, is called a **polynomial of degree n in the variable** x. The numbers $a_n, a_{n-1}, a_{n-2}, \ldots, a_1$, and a_0 are called the **coefficients** of the polynomial, and n is called its **degree**. The degree of the polynomial is the highest degree of its terms.

EXERCISE 6.6

Determine the degree of each of the following polynomials.

a) $3x^3 - 4x^2 + 5$

b) $-5x^2 + 7 - 2x^3 - 5x^4$

c) $7 - 3x^3 + 2x - x^5 - x^2$

d) $2x^3 - 3x + 4x^5 - 5x^7 + 6x^6$

e) $2x(x^4 - 5x^2 + 7x)$

f) $8 - 5x$

SOLUTION 6.6

a) The terms of the polynomial are of degrees 3, 2, and 0 (A constant term is of degree 0.) Hence, the degree of the polynomial is 3.

b) The terms of the polynomial are of degrees 2, 0, 3, and 4; the degree of the polynomial is 4.

c) The terms of the polynomial are of degrees 0, 3, 1, 5, and 2; the degree of the polynomial is 5.

d) The terms of the polynomial are of degrees 3, 1, 5, 7, and 6; the degree of the polynomial is 7.

e) $2x(x^4 - 5x^2 + 7x) = 2x^5 - 10x^3 + 14x^2$. The terms of the polynomial are of degrees 5, 3, and 2; the degree of the polynomial is 5.

f) The terms of the polynomial are of degrees 0 and 1; the degree of the polynomial is 1.

Descending and Ascending Order

Definition:

A polynomial is arranged in **descending order** if the degrees of its terms decrease from left to right. A polynomial is arranged in **ascending order** if the degrees of its terms increase from left to right.

EXERCISE 6.7

(Re)arrange each of the following polynomials in descending order.

a) $2x^2 - 3x + 7 + x^5$

b) $4x^7 - 7x^2 + x - x^4 - 8x^3$

c) $2 - 3x + 4x^2 + 5x^3 - x^4$

d) $x^5 - x^2 + 4 - 6x^4 + 2$.

SOLUTION 6.7

a) $x^5 + 2x^2 - 3x + 7$

b) $4x^7 - x^4 - 8x^3 - 7x^2 + x$

c) $-x^4 + 5x^3 + 4x^2 - 3x + 2$

d) $x^5 - 6x^4 - x^2 + 2x + 4$

EXERCISE 6.8

(Re)arrange each of the following polynomials in ascending order.

a) $3x^3 - 5x^2 + 7x + 4$

b) $5x^2 - 8x^5 + 1 - 11x^4 + x$

c) $x^2 - x^6 + 5x + 13x^4 - 9$

d) $x^5 - 7 + 13x$

SOLUTION 6.8

a) $4 + 7x - 5x^2 + 3x^3$

b) $1 + x + 5x^2 - 11x^4 - 8x^5$

c) $-9 + 5x + x^2 + 13x^4 - x^6$

d) $-7 + 13x + x^5$

Division Algorithm

Division Algorithm:

If $f(x)$ and $D(x)$ are polynomials of degrees m and n, respectively, such that $m \geq n$ and $D(x) \neq 0$, then there exist unique polynomials $Q(x)$ and $R(x)$ such that

$$f(x) = D(x) \cdot Q(x) + R(x)$$

where the degree of $R(x)$ is less than the degree of $D(x)$. ($R(x)$ could be equal to 0.) $D(x)$ is called the **divisor**; $f(x)$ is called the **dividend**; $Q(x)$ is called the **quotient**; and $R(x)$ is called the **remainder**.

EXERCISE 6.9

Let $f(x) = 4x^3 + 3x^2 - 5x - 1$ and $D(x) = x - 2$. Use long division to determine $Q(x)$ and $R(x)$ such that the Division Algorithm is satisfied.

SOLUTION 6.9

$$
\begin{array}{r}
4x^2 + 11x + 17 = Q(x) \\
x - 2 \enclose{longdiv}{4x^3 + 3x^2 - 5x - 1} \\
\underline{4x^3 - 8x^2} \\
11x^2 - 5x - 1 \\
\underline{11x^2 - 22x} \\
17x - 1 \\
\underline{17x - 34} \\
33 = R(x)
\end{array}
$$

Therefore, $Q(x) = 4x^2 + 11x + 17$ and $R(x) = 33$.

EXERCISE 6.10

Let $f(x) = 2x^4 - 4x^3 + 5x^2 - x + 7$ and $D(x) = x + 3$. Use long division to determine $Q(x)$ and $R(x)$ such that the Division Algorithm is satisfied.

SOLUTION 6.10

$$
\begin{array}{r}
2x^3 - 10x^2 + 35x - 106 = Q(x) \\
x+3 \enclose{longdiv}{2x^4 - 4x^3 + 5x^2 - x + 7} \\
\underline{2x^4 + 6x^3 } \\
-10x^3 + 5x^2 - x + 7 \\
\underline{-10x^3 - 30x^2 } \\
35x^2 - x + 7 \\
\underline{35x^2 + 105x } \\
-106x + 7 \\
\underline{-106x - 318} \\
325 = R(x)
\end{array}
$$

Therefore, $Q(x) = 2x^3 - 10x^2 + 35x - 106$ and $R(x) = 325$.

Synthetic Division

Instead of performing the long division as indicated in the preceding exercises, we could use a method known as **synthetic division**. In synthetic division, we use only the coefficients of the polynomial, with the polynomial arranged in descending order. The divisor must be of the form $x - c$.

The method of synthetic division will be illustrated using the polynomials given in Exercise 6.9. The dividend is $4x^3 + 3x^2 - 5x - 1$ and the divisor is $x - 2$. The divisor is of the form $x - c$. Instead of using -2 in the divisor, we use 2, and instead of subtracting at each stage in the long division process, we add.

Step 1:
Write the row of coefficients; leave a space below; draw a line segment as indicated; bring down the leading coefficient, 4.

$$
\begin{array}{r|rrrr}
2 & 4 & 3 & -5 & -1 \\
\hline
  & 4
\end{array}
$$

Step 2:
Multiply the 2 in the "box" by the 4 brought down; enter the 8 below the coefficient 3. Add as indicated.

$$
\begin{array}{r|rrrr}
2 & 4 & 3 & -5 & -1 \\
  &   & 8 & & \\
\hline
  & 4 & 11
\end{array}
$$

Step 3:
Multiply the result, 11, by the 2 in the "box"; obtain 22. Enter the 22 below the coefficient -5. Add as

$$
\begin{array}{r|rrrr}
2 & 4 & 3 & -5 & -1 \\
  &   & 8 & 22 & \\
\hline
  & 4 & 11 & 17
\end{array}
$$

indicated.

Step 4:

Multiply the result,17, by the 2 in the "box"; enter the 34 below the coefficient −1. Add as indicated.

$$\begin{array}{r|rrrr} 2\, | & 4 & 3 & -5 & -1 \\ & & 8 & 22 & 34 \\ \hline & 4 & 11 & 17 & 33 \end{array}$$

The first three entries in the third row are the coefficients of the polynomial $Q(x)$, and the last entry is the value of $R(x)$. Hence, $Q(x) = 4x^2 + 11x + 17$ and $R(x) = 33$. These results compare with the results obtained in Exercise 6.9.

EXERCISE 6.11

Use synthetic division to divide the polynomial $F(x) = x^2 - 4x + 2x^3 + 5$ by $x + 1$.

SOLUTION 6.11

First, we rewrite $f(x)$ in descending order as $f(x) = 2x^3 + x^2 - 4x + 5$. We place −1 in the "box" since $x + 1 = x - (-1)$.

Step 1:

Write the row of coefficients; leave a space below; draw a line segment as indicated; bring down the leading coefficient, 2.

$$\begin{array}{r|rrrr} -1\, | & 2 & 1 & -4 & 5 \\ & & & & \\ \hline & 2 \end{array}$$

Step 2:

Multiply the −1 in the "box" by the 2 brought down; enter the −2 below the coefficient 1. Add as indicated.

$$\begin{array}{r|rrrr} -1\, | & 2 & 1 & -4 & 5 \\ & & -2 & & \\ \hline & 2 & -1 \end{array}$$

Step 3:

Multiply the result, −1, by the −1 in the "box"; enter the −3 below the coefficient −4. Add as indicated.

$$\begin{array}{r|rrrr} -1\, | & 2 & 1 & -4 & 5 \\ & & -2 & 1 & \\ \hline & 2 & -1 & -3 \end{array}$$

Step 4:

Multiply the result, −3, by the −1 in the "box"; enter the 3 below the coefficient 5. Add as indicated.

$$\begin{array}{r|rrrr} -1\, | & 2 & 1 & -4 & 5 \\ & & -2 & 1 & 3 \\ \hline & 2 & -1 & -3 & 8 \end{array}$$

We now determine that $Q(x) = 2x^2 - x - 3x$ and $R(x) = 8$.

EXERCISE 6.12

Use synthetic division to divide $4x^4 - 2x^3 + x + 2$ by $x + 3$.

SOLUTION 6.12

When writing the row of coefficients, we enter 0 for the coefficient of the x^2-term.

$$\begin{array}{r|rrrrr} -3 & 4 & -2 & 0 & 1 & 2 \\ & & -12 & 42 & -126 & 375 \\ \hline & 4 & -14 & 42 & -125 & 377 \end{array}$$

Therefore, the quotient is $4x^3 - 14x^2 + 42x - 125$ and the remainder is 377.

EXERCISE 6.13

Use synthetic division to divide $4x^2 - 7 + 3x^4 - 2x^5 + x$ by $x - 2$.

SOLUTION 6.13

First, we rearrange the dividend in descending order as $-2x^5 + 3x^4 + 4x^2 + x - 7$. When writing the row of coefficients, we enter 0 for the coefficient of the x^3-term.

$$\begin{array}{r|rrrrrr} 2 & -2 & 3 & 0 & 4 & 1 & -7 \\ & & -4 & -2 & -4 & 0 & 2 \\ \hline & -2 & -1 & -2 & 0 & 1 & -5 \end{array}$$

Therefore, the quotient is $-2x^4 - x^3 - 2x^2 + 1$ and the remainder is -5.

Remainder Theorem

Using synthetic division, we can evaluate polynomials for certain real number values. For instance, in Exercise 6.13 we determined that
$$f(x) = -2x^5 + 3x^4 + 4x^2 + x - 7 = (-2x^4 - x^3 - 2x^2 + 1)(x - 2) - 5$$
Also,
$$\begin{aligned} f(2) &= [-2(2)^4 - (2)^3 - 2(2)^2 + 1](2 - 2) - 5 \\ &= [-2(2)^4 - (2)^3 - 2(2)^2 + 1](0) - 5 \\ &= 0 - 5 \\ &= -5. \end{aligned}$$

Hence, $f(2) = -5 = R(x)$, the remainder in the synthetic division problem. Since $R(x)$ is a constant, we can refer to it simply as R. Therefore, we have $f(2) = -5 = R$.

> **Remainder Theorem:**
> If $f(x)$ is a polynomial of degree greater than or equal to 1, and R is the remainder when $f(x)$ is divided by $x - c$, where c is any real number, then $f(c) = R$.

The proof of the Remainder Theorem follows. By the Division Algorithm, we have

$$f(x) = Q(x) \cdot (x - c) + R$$

which is true for all real values of x. In particular, if $x = c$, then

$$f(c) = Q(c) \cdot (c - c) + R$$

$$= Q(c) \cdot (0) + R$$

$$= 0 + R$$

$$= R$$

Hence, $f(c) = R$.

EXERCISE 6.14

Determine the value of each of the following, using synthetic division and the Remainder Theorem.

a) $f(2)$, if $f(x) = 2x^3 - 3x^2 + x - 4$

b) $f(-3)$, if $f(x) = x^4 - 3x^3 + 7x - 1$

SOLUTION 6.14

a) By the Remainder Theorem, $f(2)$ is the remainder when $f(x)$ is divided by $(x - 2)$. Using synthetic division, we have

$$
\begin{array}{r|rrrr}
2\underline{)} & 2 & -3 & 1 & -4 \\
 & & 4 & 2 & 6 \\
\hline
 & 2 & 1 & 3 & 2
\end{array}
$$

The remainder is 2; hence, $f(2) = 2$.

b) By the Remainder Theorem, $f(-3)$ is the remainder when $f(x)$ is divided by $(x + 3)$. Using synthetic division, we have

$$
\begin{array}{r|rrrrr}
-3\underline{)} & 1 & -3 & 0 & 7 & -1 \\
 & & -3 & 18 & -54 & 141 \\
\hline
 & 1 & -6 & 18 & -47 & 140
\end{array}
$$

The remainder is 140; hence, $f(-3) = 140$.

Factor Theorem

It follows from the above, that if $f(c) = R = 0$, then both $x - c$ and $Q(x)$ are factors of $f(x)$.

> **Factor Theorem:**
>
> Let $f(x)$ be a polynomial of degree greater than or equal to 1, and let R be the remainder when $f(x)$ is divided by $x - c$, where c is any real number. If $R = 0$, then $x - c$ is a factor of $f(x)$ and, conversely, if $x - c$ is a factor of $f(x)$, then $r = 0$.

EXERCISE 6.15

Use the Factor Theorem to determine whether

a) $x - 2$ is a factor of $f(x) = x^3 + x^2 - 4x - 7$

b) $x + 3$ is a factor of $f(x) = 2x^4 + 5x^3 - x^2 + 5x - 3$

SOLUTION 6.15

a) The expression $x - 2$ is of the form $x - c$ with $c = 2$. Using synthetic division, we determine the remainder when $f(x)$ is divided by $x - 2$.

$$
\begin{array}{r|rrrr}
2\rfloor & 1 & 1 & -4 & -7 \\
& & 2 & 6 & 4 \\
\hline
& 1 & 3 & 2 & -3 = R
\end{array}
$$

Since $R \neq 0$, then, by the Factor Theorem, $x - 2$ is *not* a factor of $f(x)$.

b) The expression $x + 3$ is of the form $x - c$ with $c = -3$. Using synthetic division, we determine the remainder when $f(x)$ is divided by $x + 3$.

$$
\begin{array}{r|rrrrr}
-3\rfloor & 2 & 5 & -1 & 5 & -3 \\
& & -6 & 3 & -6 & 3 \\
\hline
& 2 & -1 & 2 & -1 & 0 = R
\end{array}
$$

Since $R = 0$, then, by the Factor Theorem, $x + 3$ is a factor of $f(x)$.

6.3 REAL ZEROS OF HIGHER-DEGREE POLYNOMIALS

The graphs of general polynomial functions of degree $n > 2$ are not as easy to obtain as are the graphs of linear, quadratic, and power functions.

To graph a general polynomial function, it is very helpful to determine the real zeros of the polynomial. Recall that each real zero of a polynomial corresponds to the x-coordinate of a point where the graph of the polynomial function crosses or is tangent to the x-axis. The set of *all* of the real zeros of the polynomial will thus partition, or subdivide, the domain of the polynomial function into disjoint subsets. On each such subset of the domain, the function values will all be of the same algebraic sign. Hence, it can be determined whether the graph is above or below the x-axis on a particular subset of the domain.

To obtain the real zeros of a general polynomial, we will use the techniques discussed in the previous section of this chapter. Other techniques will be discussed in this chapter. In a calculus course, you will learn additional techniques.

6.4 NUMBER OF ZEROS OF A POLYNOMIAL

Rule:

Let f be a polynomial function in the variable x, of degree n. Then, $f(x)$ cannot have more than n zeros.

The above statement follows from the Factor Theorem. Each zero, c, of the polynomial corresponds to a linear factor, $x - c$. Further, an nth-degree polynomial cannot have more than n linear factors.

The zeros of a polynomial do not have to be distinct. That is, some zeros may be repeated. Some zeros may not be real, as we have seen in the case of quadratic polynomials.

Definition:

A zero that occurs twice is called a **double zero**. A zero that occurs three times is called a **triple zero**. In general, a zero that occurs k times, is called **a zero of multiplicity k**.

Rule:

Let $f(x)$ be a polynomial of degree $n \geq 1$. If all zeros of multiplicity k are counted as k zeros, and if all nonreal zeros are also counted, then $f(x)$ has *exactly n zeros*.

EXERCISE 6.16

Use synthetic division to determine all of the real zeros of the polynomial $f(x) = x^3 - 4x^2 + x + 6$.

SOLUTION 6.16

Using trial and error, we try any arbitrary real value for a zero of $f(x)$, since $f(x)$ is defined for all real values of x. For instance, we can try $x = 1$ and proceed with synthetic division.

$$
\begin{array}{r|rrrr}
1 & 1 & -4 & 1 & 6 \\
  &   & 1 & -3 & -2 \\
\hline
  & 1 & -3 & -2 & 4 = R
\end{array}
$$

For 1 to be a zero of $f(x)$, R must be 0. Since $R = 4 \neq 0$, 1 is not a zero of $f(x)$. We will try -1.

$$
\begin{array}{r|rrrr}
-1 & 1 & -4 & 1 & 6 \\
   &   & -1 & 5 & -6 \\
\hline
   & 1 & -5 & 6 & 0 = R
\end{array}
$$

Since $R = 0$, -1 is a zero of $f(x)$. Further, $x - (-1) = x + 1$ is a factor of $f(x)$. The first three entries in the last line of the above synthetic division are the coefficients for the other factor of $f(x)$. Hence,

$$f(x) = (x + 1)(x^2 - 5x + 6).$$

The expression $x^2 - 5x + 6$ is factorable as $(x - 2)(x - 3)$. We have
$$f(x) = (x + 1)(x - 2)(x - 3).$$
Since $(x + 1)$, $(x - 2)$, and $(x - 3)$ are factors of $f(x)$, then -1, 2, and 3 are zeros of $f(x)$. Therefore, the real zeros of the given polynomial are -1, 2, and 3.

Rational Zeros

Instead of attempting to determine the zeros of a polynomial by trial and error, we have a more systematic procedure, which restricts our attention to real zeros that are rational. A **rational zero** is a real zero that can be expressed as the quotient of two integers such that the denominator is a *positive* integer.

> **Rational Zero Theorem:**
>
> If $f(x) = a_n x^n + a_{n-1} x^{n-1} + \cdots + a_1 x + a_0$ is an nth-degree polynomial in the variable x, such that all of the coefficients are *integers*, then any *rational* zero of $f(x)$ can be expressed in the form $\dfrac{p}{q}$, simplified to lowest terms, where p is an *integer* factor of a_0 and q is a *positive integer* factor of a_n.

EXERCISE 6.17

Using the Rational Zero Theorem, determine all of the rational numbers that may be considered as possible zeros of the polynomial $f(x) = 2x^4 - 5x^3 + x - 4$.

SOLUTION 6.17

The coefficients of $f(x)$ are integers with $a_n = 2$ and $a_0 = -4$. The integer factors of $a_0 = -4$ are ± 1, ± 2, and ± 4. The positive integer factors of $a_n = 2$ are 1 and 2. We now form the rational numbers of the form $\dfrac{p}{q}$ such that p is an integer factor of -4 and q is a positive integer factor of 2. We have

$$\frac{\pm 1}{1}, \frac{\pm 2}{1}, \frac{\pm 4}{1}, \frac{\pm 1}{2}, \frac{\pm 2}{2}, \frac{\pm 4}{2}.$$

Simplifying each of these gives the largest possible subset of R of "candidates" for rational zeros of $f(x)$. They are

$$\frac{\pm 1}{2}, \pm 1, \pm 2, \pm 4.$$

EXERCISE 6.18

Determine all rational zeros of the polynomial $f(x) = 2x^4 + 7x^2 - 9$.

SOLUTION 6.18

The coefficients of $f(x)$ are all integers, $a_n = 2$, and $a_0 = -9$. The integer factors of -9 are ± 1, ± 3, and ± 9. The positive integer factors of 2 are 1 and 2. Hence, the "candidates" for rational zeros of $f(x)$ are

$$\frac{\pm 1}{1}, \frac{\pm 3}{1}, \frac{\pm 9}{1}, \frac{\pm 1}{2}, \frac{\pm 3}{2}, \frac{\pm 9}{2}$$

which simplify to

$$\frac{\pm 1}{2}, \pm 1, \frac{\pm 3}{2}, \pm 3, \frac{\pm 9}{2}, \pm 9.$$

Using synthetic division and noting that the coefficients of the third degree and first degree terms of $f(x)$ are 0, we test to see which of the "candidates" are actually zeros of $f(x)$.

$$
\begin{array}{r|rrrrr}
\frac{-1}{2} & 2 & 0 & 7 & 0 & -9 \\
& & -1 & \frac{1}{2} & \frac{-15}{4} & \frac{15}{8} \\
\hline
& 2 & -1 & \frac{15}{2} & \frac{-15}{4} & \frac{-57}{8} = R
\end{array}
\qquad
\begin{array}{r|rrrrr}
\frac{1}{2} & 2 & 0 & 7 & 0 & -9 \\
& & 1 & \frac{1}{2} & \frac{15}{4} & \frac{15}{8} \\
\hline
& 2 & 1 & \frac{15}{2} & \frac{15}{4} & \frac{-57}{8} = R
\end{array}
$$

Since $R = \dfrac{-57}{8} \neq 0$, $\dfrac{-1}{2}$ is **not** a zero of $f(x)$. Since $R = \dfrac{-57}{8} \neq 0$, $\dfrac{1}{2}$ is **not** a zero of $f(x)$.

$$
\begin{array}{r|rrrrr}
-1 & 2 & 0 & 7 & 0 & -9 \\
& & -2 & 2 & -9 & 9 \\
\hline
& 2 & -2 & 9 & -9 & 0 = R
\end{array}
\qquad
\begin{array}{r|rrrrr}
1 & 2 & 0 & 7 & 0 & -9 \\
& & 2 & 2 & 9 & 9 \\
\hline
& 2 & 2 & 9 & 9 & 0 = R
\end{array}
$$

Since $R = 0$, -1 is a zero of $f(x)$. Since $R = 0$, -1 **is** a zero of $f(x)$.

Since -1 is a zero of $f(x)$, then $x + 1$ is a factor of $f(x)$. Using the synthetic division above, we determine that the other factor of $f(x)$ is $2x^3 - 2x^2 + 9x - 9$. But, 1 is also a zero of $f(x)$ and, therefore, is a zero of $2x^3 - 2x^2 + 9x - 9$. Using synthetic division again, we have

$$
\begin{array}{r|rrrr}
1 & 2 & -2 & 9 & -9 \\
& & 2 & 0 & 9 \\
\hline
& 2 & 0 & 9 & 0 = R
\end{array}
$$

Hence,

$$
\begin{aligned}
f(x) &= (x + 1)(2x^3 - 2x^2 + 9x - 9) \\
&= (x + 1)(x - 1)(2x^2 + 9).
\end{aligned}
$$

But, $2x^2 + 9$ is not factorable on the set of rational numbers. Therefore, the only rational zeros of $f(x)$ are -1 and 1.

Descartes' Rule of Signs

Determining the rational zeros of polynomials involves quite a bit of work. If the coefficients a_n and a_0 have many factors, there may be large numbers of "candidates" to test. It is therefore desirable to be able to eliminate some of these rational numbers from further consideration. The following may enable us to do this.

> **Descartes' Rule of Signs:**
>
> Let $f(x) = a_n x^n + a_{n-1} x^{n-1} + \cdots + a_1 x + a_0$ $(a_n \neq 0)$ be a polynomial with its terms arranged either in descending or ascending order.
>
> i. The maximum number of **positive** zeros of $f(x)$ is equal to the number of sign changes that occur between successive coefficients of $f(x)$. If $f(x)$ does not have the maximum number of positive zeros, then the number of positive zeros will differ from the maximum number by an *even* number.
>
> ii. The maximum number of **negative** zeros of $f(x)$ is equal to the number of sign changes that occur between successive coefficients of $f(-x)$. If $f(x)$ does not have the maximum number of negative zeros, then the number of negative zeros will differ from the maximum number by an *even* number.

EXERCISE 6.19

Determine the possible number of positive zeros and the possible number of negative zeros for the polynomial $f(x) = x^5 - 3x^4 - 2x^3 + x^2 - 3x + 7$.

SOLUTION 6.19

$$f(x) = x^5 - 3x^4 - 2x^3 + x^2 - 3x + 7$$

$+$ to $-$ $-$ to $+$ $+$ to $-$ $-$ to $+$

The polynomial $f(x)$ is arranged in descending order. There are four sign changes between successive terms. Therefore, the *maximum* number of positive zeros is four. However, if four is not the number of positive zeros, the required number must be $4 - 2 = 2$, or $4 - 4 = 0$. Hence, there are four, two, or no positive zeros for $f(x)$.

To determine the number of negative zeros for $f(x)$, we must consider $f(-x)$.

$$f(-x) = (-x)^5 - 3(-x)^4 - 2(-x)^3 + (-x)^2 - 3(-x) + 7$$
$$= -x^5 - 3x^4 + 2x^3 + x^2 + 3x + 7$$

$-$ to $+$

There is only one sign change in $f(x)$. Therefore, the *maximum* number of negative zeros of $f(x)$ is one. Since $1 - 2 < 0$, there will be exactly one negative zero for $f(x)$.

EXERCISE 6.20

Determine all rational zeros for the polynomial $f(x) = 2x^4 - x^3 - 14x^2 - 5x + 6$.

SOLUTION 6.20

Testing for possible numbers of positive and negative zeros, we note that $f(x)$ is arranged in descending order and has two sign changes between successive terms. Therefore, $f(x)$ has either two or no positive zeros. Checking for negative zeros, we have

$$f(-x) = 2x^4 + x^3 - 14x^2 + 5x + 6$$

which is arranged in descending order and has two sign changes between successive terms. Therefore, $f(x)$ has either two or no negative zeros.

The integer factors of $a_0 = 6$ are ± 1, ± 2, ± 3, and ± 6. The positive integer factors of $a_n = 2$ are 1 and 2. The "candidates" for rational zeros, then, are

$$\frac{\pm 1}{1}, \frac{\pm 2}{1}, \frac{\pm 3}{1}, \frac{\pm 6}{1}, \frac{\pm 1}{2}, \frac{\pm 2}{2}, \frac{\pm 3}{2}, \frac{\pm 6}{2}$$

which simplify to

$$\pm 1, \pm 2, \pm 3, \pm 6, \frac{\pm 1}{2}, \frac{\pm 3}{2}.$$

Using synthetic division, we have

1|	2	−1	−14	−5	6
			2	1	−13	−18
		2	1	−13	−18	−12 = R

−1|	2	−1	−14	−5	6
			−2	3	11	−6
		2	−3	−11	6	0 = R

We conclude that −1 is a real zero of $f(x)$ but 1 is not. At this point, we have found one negative zero. Therefore, there must be another negative zero. We continue.

2|	2	−1	−14	−5	6
			4	6	−16	−42
		2	3	−8	−21	−36 = R

−2|	2	−1	−14	−5	6
			−4	10	8	−6
		2	−5	−4	3	0 = R

We conclude that −2 is a real zero of $f(x)$ but 2 is not. At this point, we have found two negative zeros. Since two is the maximum number of negative zeros, we shall concentrate on positive zeros only. We continue.

3|	2	−1	−14	−5	6
			6	15	3	−6
		2	5	1	−2	0 = R

6|	2	−1	−14	−5	6
			12	66	312	1842
		2	11	52	307	1848 = R

We conclude that 3 is a real zero of $f(x)$ but 6 is not. We now have one positive zero and will look for a second one. We continue.

$$\frac{1}{2} \bigg| \begin{array}{ccccc} 2 & -1 & -14 & -5 & 6 \\ & 1 & 0 & -7 & -6 \\ \hline 2 & 0 & -14 & -12 & 0 = R \end{array}$$

We conclude that $\frac{1}{2}$ is a zero of $f(x)$. We have now determined two positive zeros for $f(x)$. Since two is the maximum number of positive zeros for $f(x)$, we are done. Hence, the rational zeros of $f(x)$ are $-2, -1, \frac{1}{2}$, and 3.

6.5 IRRATIONAL ZEROS OF POLYNOMIALS

We have considered the problem of determining real zeros of polynomials. If the polynomial has rational zeros, we are able to determine what they are by the use of the Rational Zero Theorem. There are polynomials, however, with real zeros that are not rational. A real zero that is not rational is irrational. The following theorem will be helpful in determining irrational zeros.

Intermediate Value Theorem

Intermediate Value Theorem:
Let $f(x)$ be an nth-degree polynomial in the variable x. Let x_1 and x_2 be values in the domain of f such that $x_1 \neq x_2$. If b is a real value between $f(x_1)$ and $f(x_2)$, then there exists a real value c between x_1 and x_2 such that $b = f(c)$.

The geometric significance of this theorem is illustrated in Figure 6.9. The given polynomial is $f(x) = x^3$. For the values x_1 and x_2, we located $f(x_1)$ and $f(x_2)$ as indicated. Next, a value b was chosen between $f(x_1)$ and $f(x_2)$ as indicated. A line segment was constructed through the point $(0, b)$ parallel to the x-axis and intersecting the graph at the point R. Through the point R, a line segment was constructed perpendicular to the x-axis. The x-coordinate of the point of intersection of this line segment with the x-axis is the required value c of the theorem. Clearly, $b = f(c)$.

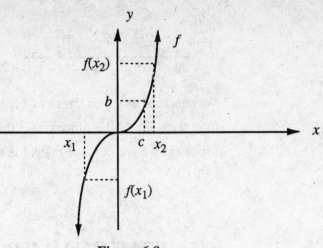

Figure 6.9
The Graph of $f(x) = x^3$

It should be noted that the Intermediate Value Theorem is valid for functions that are continuous. Intuitively, a function is continuous if its graph can be traced from one end of its domain to the other without lifting the pencil from the paper; that is, there are no "holes" in the graph of a continuous function. (A precise definition of continuity is given in a beginning calculus course.) All polynomial functions are continuous on their domains. Hence, the theorem applies to all polynomials.

We are attempting to determine irrational zeros of a polynomial. If such zeros exist, we would bracket each by locating a value, x_1, such that $f(x_1) > 0$, and another value, x_2, such that $f(x_2) < 0$. Then, by the Intermediate Value Theorem, there would exist a real value c between x_1 and x_2 such that $f(c) = 0$. The value c would be a zero of the polynomial.

EXERCISE 6.21

If there exists a zero of the polynomial $f(x) = x^2 - 3$ between 1 and 2, determine its value, correct to two decimal places.

SOLUTION 6.21

$f(1) = (1)^2 - 3 = -2$ and $f(2) = (2)^2 - 3 = 1$. Since $f(1)$ and $f(2)$ have opposite signs, there exists a zero between 1 and 2. The next step would be to bracket the zero between numbers correct to the nearest tenth. Taking 1.5 as the average of 1 and 2, we determine that $f(1.5) = (1.5)^2 - 3 = -0.75$ and that the zero is between 1.5 and 2. To the nearest tenth, the average of 1.5 and 2 is 1.8. $f(1.8) = (1.8)^2 - 3 = 0.24$. The zero is between 1.5 and 1.8. Proceeding in a

similar manner, we determine that $f(1.7) = (1.7)^2 - 3 = -0.11$.
Hence, the zero is between 1.7 and 1.8. For 1.75, we have
$f(1.75) = (1.75)^2 - 3 = 0.0626$. The zero is between 1.7 and 1.75.
$f(1.73) = (1.73)^2 - 3 = -0.0071$. The zero is between 1.73 and 1.75.
$f(1.74) = (1.74)^2 - 3 = 0.0276$. The zero is between 1.73 and 1.74.
$f(1.735) = (1.735)^2 - 3 = 0.010225$. The zero is between 1.73 and
1.735. Rounded to the nearest tenth, the zero of $f(x)$ is 1.73.

In Figure 6.10, we graph the function $f(x) = x^2 - 3$ and observe that
the graph crosses the x-axis at approximately $x = 1.73$.

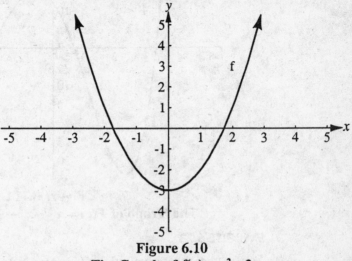

Figure 6.10
The Graph of $f(x) = x^2 - 3$

EXERCISE 6.22

If the polynomial $f(x) = x^4 - x^3 - 11x^2 + 12x + 12$ has a negative zero,
determine its value, correct to the nearest tenth.

SOLUTION 6.22

We will start our search between −1 and 0. Using a calculator, we determine values of $f(x)$.

$f(0) = 12$ and $f(-1) = -9$

Hence, there exists a real zero between −1 and 0.

$f(-0.5) = 3.4375$

Hence, there exists a real zero between −1 and −0.5.

$f(-0.8) = -3.7184$

Hence, there exists a real zero between −0.8 and −0.5.

$f(-0.6) = 1.1856$

Hence, there exists a real zero between −0.8 and −0.6.

$f(-0.7) = -1.2069$

Hence, there exists a real zero between −0.7 and −0.6.

$$f(-0.65) = 0.005\overset{.}{6}315$$

Hence, there exists a real zero between –0.7 and –0.65.

Therefore, a negative zero does exist for $f(x)$ at –0.7, rounded to the nearest tenth.

In Figure 6.11, we graph the function $f(x) = x^4 - x^3 - 11x^2 + 12x + 12$ and observe that the graph crosses the x-axis at approximately $x = -0.7$.

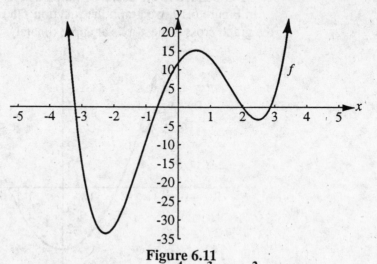

Figure 6.11

The Graph of $F(x) = x^4 - x^3 - 11x^2 + 12x + 12$

EXERCISE 6.23

Verify that the polynomial $f(x) = x^4 + 2x^3 - 6x^2 - 4x + 8$ has exactly two positive zeros and that both of these zeros lie between $x = 1$ and $x = 2$.

SOLUTION 6.23

We note that $f(x)$ is arranged in descending order and has two sign changes between successive terms. Therefore, according to Descartes' Rule, $f(x)$ has either two or no positive zeros. Using the Intermediate Value Theorem and a calculator, we determine that

$$f(1) = 1 \qquad \text{and} \qquad f(2) = 8$$

Since $f(1)$ and $f(2)$ have the same sign, there may or may not be a zero between 1 and 2. We compute $f(1.5)$.

$$f(1.5) = 0.3125$$

Again, it appears that there may not be a zero between 1 and 2 since $f(1)$, $f(1.5)$, and $f(2)$ all have the same algebraic sign. We compute $f(1.4)$.

$$f(1.4) = -0.0304$$

Since $f(1)$ and $f(1.4)$ have opposite signs, there is a real zero between 1 and 1.4. Since there is one positive zero, there has to be another. Since $f(1.4)$ and $f(2)$ have opposite signs, there is also a real zero between 1.4

and 2. Hence, there are exactly two positive zeros and they are both between 1 and 2. (See Figure 6.12.)

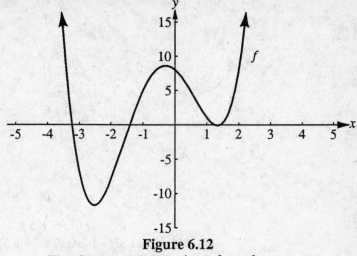

Figure 6.12
The Graph of $f(x) = x^4 + 2x^3 - 6x^2 - 4x + 8$

6.5 GRAPHS OF POLYNOMIAL FUNCTIONS

To graph higher-degree polynomial functions, we try to determine all of the real zeros of the polynomial. This is done using the techniques introduced so far in this chapter.

EXERCISE 6.24

Determine the graph of the polynomial function A defined by $A(x) = x^3 - 7x + 6$.

SOLUTION 6.24

$A(x)$ is a polynomial arranged in descending order. $A(x)$ has two sign changes between successive terms. Hence, $A(x)$ has either two or no positive zeros. Further, $A(-x) = -x^3 + 7x + 6$ has one sign change between successive terms. Therefore, $A(x)$ has exactly one negative zero. Using the Rational Zero Theorem and synthetic division, we determine that -3, 1, and 2 are real zeros of $A(x)$. Therefore, $(x+3)$, $(x-1)$, and $(x-2)$ are factors of $A(x)$. Hence, $A(x) = (x+3)(x-1)(x-2)$. The three real zeros partition, or subdivide, the entire set of real numbers into the subsets $(-\infty, -3)$, $(-3, 1)$, $(1, 2)$, and $(2, +\infty)$. (See Figure 6.13.) On each of these intervals, $A(x)$ will have the same sign for all values of x in the interval. We now proceed to test for the algebraic sign of $A(x)$ on each of the intervals.

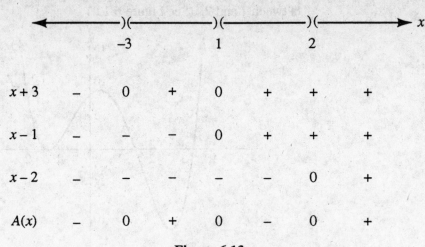

Figure 6.13
Algebraic Sign of $A(x) = (x + 3)(x - 1)(x - 2)$

We determine that $A(x) > 0$ when x is in the intervals $(-3, 1)$ and $(2, +\infty)$. $A(x) < 0$ when x is in the intervals $(-\infty, -3)$ and $(1, 2)$. Using a graphing calculator, we determine the graph of A. (See Figure 6.14.)

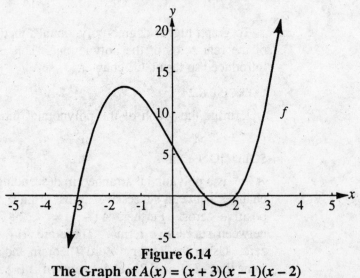

Figure 6.14
The Graph of $A(x) = (x + 3)(x - 1)(x - 2)$

EXERCISE 6.25

Graph the function B defined by $B(x) = x^4 - 2x^2 + 1$.

SOLUTION 6.25

Since $a_1 = 1$ and $a_0 = 1$, the only "candidates" for rational zeros of $B(x)$ are -1 and 1. Using synthetic division, we determine that 1 is a *double* zero and that -1 is also a *double* zero. Hence, $B(x) = (x-1)^2 (x+1)^2$. We now proceed to test for the algebraic sign for $B(x)$ on each of the intervals determined by the real zeros -1 and 1. (See Figure 6.15.)

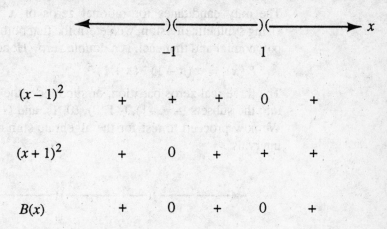

Figure 6.15
Algebraic Sign of $B(x) = x^4 - 2x^2 + 1$

We determine that $B(x) > 0$ for all real values of x except $x = -1$ and $x = 1$. Using a graphing calculator, we determine the graph of B. (See Figure 6.16.) Note that B is an even function; its graph is symmetric with respect to the y-axis.

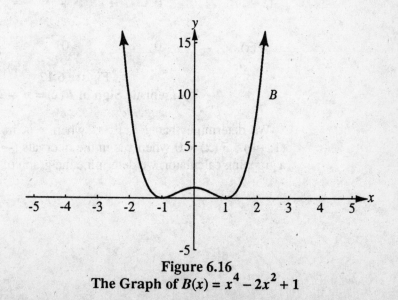

Figure 6.16
The Graph of $B(x) = x^4 - 2x^2 + 1$

EXERCISE 6.26

Graph the function F defined by $F(x) = x^5 - 2x^3 + x$.

SOLUTION 6.26

Since $F(x)$ contains no constant term, 0 is a zero of $F(x)$. Hence,

$$F(x) = x(x^4 - 2x^2 + 1).$$

The only candidates for rational zeros of $x^4 - 2x^2 + 1$ are -1 and 1. Using synthetic division, we determine that both -1 and 1 are zeros of the polynomial and that each is a double zero. Hence,

$$F(x) = x(x-1)^2(x+1)^2.$$

The three real zeros partition, or subdivide, the entire set of real numbers into the subsets $(-\infty, -1)$, $(-1, 0)$, $(0, 1)$, and $(1, +\infty)$. (See Figure 6.17.) We now proceed to test for the algebraic sign of $F(x)$ on each of these intervals.

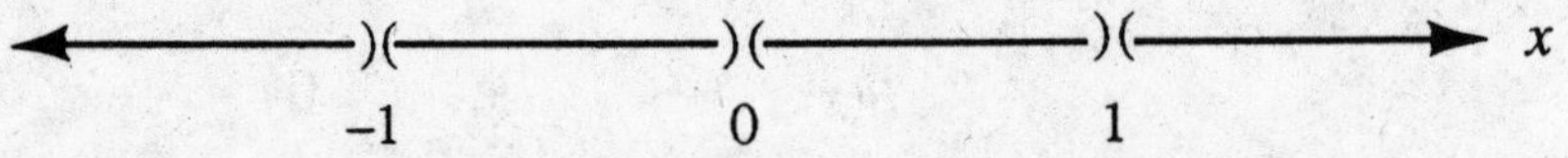

	$(-\infty,-1)$	-1	$(-1,0)$	0	$(0,1)$	1	$(1,+\infty)$
x	$-$	$-$	$-$	0	$+$	$+$	$+$
$(x-1)^2$	$+$	$+$	$+$	$+$	$+$	0	$+$
$(x+1)^2$	$+$	0	$+$	$+$	$+$	$+$	$+$
$F(x)$	$-$	0	$-$	0	$+$	0	$+$

Figure 6.17

Algebraic Sign of $F(x) = x^5 - 2x^2 + x$

We determine that $F(x) > 0$ when x is in the intervals $(0, 1)$ and $(1, +\infty)$. $F(x) < 0$ when x is in the intervals $(-\infty, -1)$ and $(-1, 0)$. Using a graphing calculator, we determine the graph of F. (See Figure 6.18.)

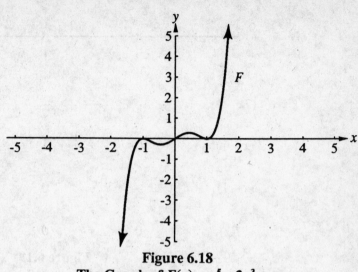

Figure 6.18
The Graph of $F(x) = x^5 - 2x^3 + x$

EXERCISE 6.27

Graph the function H defined by $H(x) = 2x^4 - x^3 - 5x^2 + x + 3$.

SOLUTION 6.27

$H(x)$ is a polynomial which is arranged in descending order. $H(x)$ has two sign changes between successive terms. According to Descartes' Rule, $H(x)$ has either two or no positive zeros. $H(-x) = 2x^4 + x^3 - 5x^2 - x + 3$ has two sign changes between successive terms. Hence, $H(x)$ has either two or no negative zeros. Using the Rational Zero Theorem, we determine that the "candidates" for rational zeros of $H(x)$ are ± 1, $\dfrac{\pm 1}{2}$, ± 3, and $\dfrac{\pm 3}{2}$. Using synthetic division, we determine that -1 is a double zero of $H(x)$. The other real zeros are 1 and $\dfrac{3}{2}$. Hence,

$$H(x) = 2(x+1)^2(x-1)\left(x - \frac{3}{2}\right)$$

or

$$H(x) = (2x-3)(x+1)^2(x-1).$$

These real zeros partition, or subdivide, the entire set of real numbers into the intervals $(-\infty, -1)$, $(-1, 1)$, $\left(1, \dfrac{3}{2}\right)$, and $\left(\dfrac{3}{2}, +\infty\right)$. We now proceed to test for the algebraic sign for $H(x)$ on each of these intervals. (See Figure 6.19.)

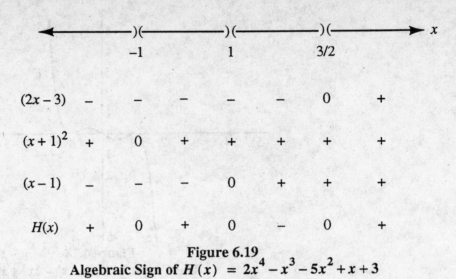

$(2x - 3)$	–	–	–	–	–	0	+
$(x + 1)^2$	+	0	+	+	+	+	+
$(x - 1)$	–	–	–	0	+	+	+
$H(x)$	+	0	+	0	–	0	+

Figure 6.19

Algebraic Sign of $H(x) = 2x^4 - x^3 - 5x^2 + x + 3$

We determine that $H(x) > 0$ when x is in the intervals $(-\infty, -1)$, $(-1, 1)$, and $\left(\dfrac{3}{2}, +\infty\right)$. $H(x) < 0$ when x is in the interval $\left(1, \dfrac{3}{2}\right)$. Using a graphing calculator, we determine the graph of H. (See Figure 6.20.)

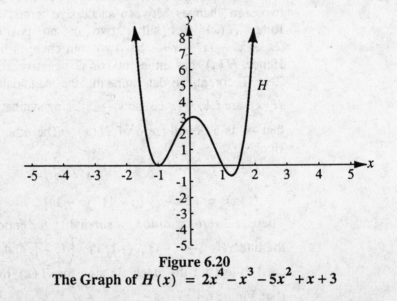

Figure 6.20

The Graph of $H(x) = 2x^4 - x^3 - 5x^2 + x + 3$

6.7 COMPLEX ZEROS OF POLYNOMIALS

So far in this chapter, we have been dealing with real zeros of polynomials. For some polynomials, complex zeros also exist. Recall from algebra that a complex number is an expression of the form $a + bi$, where a and b are real numbers and i is the imaginary unit defined by $i = \sqrt{-1}$. Complex numbers show up in the solution of quadratic equations. For instance, the equation $x^2 + 1 = 0$ or, equivalently, $x^2 = -1$, has the two complex-number solutions $\pm\sqrt{-1}$, or $\pm i$.

We observed that every linear polynomial with real coefficients has exactly one real zero. We also observed that every quadratic polynomial with real coefficients has its zeros found among the complex numbers. What can we say about the zeros of higher-degree polynomials with real coefficients? The answer is given in the following theorem.

The Fundamental Theorem of Algebra

> **The Fundamental Theorem of Algebra:**
> Every polynomial of degree $n \geq 1$ has a zero found among the complex numbers.

The Linear Factorization Theorem

The Factor Theorem also pertains to complex zeros. That is, if x_0 is a complex zero of the polynomial $f(x)$, then $x - x_0$ is a factor of $f(x)$. If $f(x)$ is of degree $n \geq 1$, then $f(x)$ has exactly n zeros, counting those that are repeated zeros. Hence, we have the following theorem.

> **The Linear Factorization Theorem:**
> Let $f(x) = a_n x^n + a_{n-1} x^{n-1} + a_{n-2} x^{n-2} + \cdots + a_1 x + a_0$ be a polynomial of degree $n \geq 1$, with real coefficients and such that $a_n \neq 0$. Then,
> $$f(x) = a(x - c_1)(x - c_2) \cdots (x - c_n)$$
> where $c_1, c_2, \ldots, c_n$ are complex numbers, not necessarily distinct.

Conjugate Zero Theorem

Recall from algebra that if $z = a + bi$ is a complex number, then its conjugate, denoted by $\overline{z}$, is $\overline{z} = a - bi$.

EXERCISE 6.28

Determine the conjugate of each of the following.

a) $3 - 2i$

b) $-2 + 5i$

c) $4 + 9i$

d) $-6 - 8i$

e) $7i$

f) 10

g) $-\pi i$

h) 1.5

SOLUTION 6.28

a) $3 + 2i$

b) $-2 - 5i$

c) $4 - 9i$

d) $-6 + 8i$

e) $-7i$

f) 10

g) πi

h) 1.5

Conjugate Zero Theorem:

Let $f(x)$ be a polynomial of degree $n > 1$ with *real* coefficients. If $a + bi$ is a zero of $f(x)$, then $a - bi$ is also a zero of $f(x)$.

EXERCISE 6.29

Form the polynomial $f(x)$ of degree 4, with real coefficients, if -1, 3, and $3 - 2i$ are zeros of $f(x)$.

SOLUTION 6.29

Since $3 - 2i$ is a zero of $f(x)$, then, from the Conjugate Zero Theorem, $3 + 2i$ is also a zero since $f(x)$ is to have real coefficients. By the Linear Factorization Theorem, we have

$$
\begin{aligned}
f(x) &= [x - (-1)] \, (x - 3) \, [x - (3 - 2i)] \, [x - (3 + 2i)] \\
&= (x + 1) \, (x - 3) \, [(x - 3) + 2i] \, [(x - 3) - 2i] \\
&= (x^2 - 2x - 3) \, [(x - 3) + 2i] \, [(x - 3) - 2i] \\
&= (x^2 - 2x - 3) \, [(x - 3)^2 - 4i^2] \\
&= (x^2 - 2x - 3) \, [(x - 3)^2 + 4] \qquad \text{(Since } i^2 = -1.)
\end{aligned}
$$

$$= x^4 - 8x^3 + 22x^2 - 8x - 39$$

Therefore,

$$f(x) = x^4 - 8x^3 + 22x^2 - 8x - 39.$$

As we noted earlier, a real zero of a polynomial corresponds to the x-coordinate of the point where the graph of the associated polynomial function crosses or is tangent to the x-axis. If a polynomial has no real zeros, then the graph of the associated polynomial function will be completely above or completely below the x-axis.

EXERCISE 6.30

Graph the function F defined by $F(x) = x^2 + 2$.

SOLUTION 6.30

The polynomial $x^2 + 2$ has exactly two zeros, both of which are complex. They are $\pm i\sqrt{2}$. Since the zeros are not real , the graph of F will not cross or be tangent to the x-axis anywhere. The graph of F is completely above the x-axis; the range of F is $[2, +\infty)$. (See Figure 6.21.)

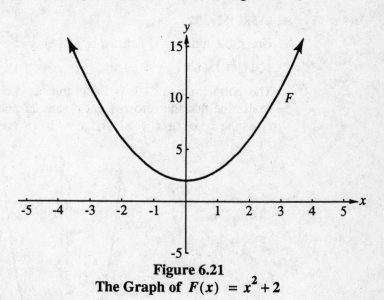

Figure 6.21
The Graph of $F(x) = x^2 + 2$

EXERCISE 6.31

Graph the function $H(x) = x^3 + 1$.

SOLUTION 6.31

The polynomial $x^3 + 1$ has exactly three zeros. One is real, namely, -1.

The other two are the conjugate complex zeros, $\dfrac{1 + i\sqrt{3}}{2}$ and $\dfrac{1 - i\sqrt{3}}{2}$.

The graph of H will cross the x-axis only at $x = -1$. (See Figure 6.22.)

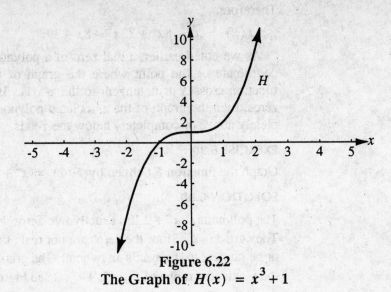

Figure 6.22
The Graph of $H(x) = x^3 + 1$

EXERCISE 6.32

Graph the function Q defined by $Q(x) = x^4 - 1$.

SOLUTION 6.32

The polynomial $x^4 - 1$ is factorable as $(x^2 + 1)(x - 1)(x + 1)$. We determine that the zeros of $Q(x)$ are $\pm i$ and ± 1. The graph of Q will cross the x-axis at $x = -1$ and $x = 1$. (See Figure 6.23.)

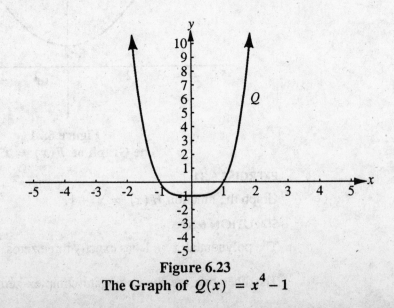

Figure 6.23
The Graph of $Q(x) = x^4 - 1$

6.8 RATIONAL FUNCTIONS

Definition:

A function f of the form

$$f(x) = \frac{p(x)}{q(x)},$$

where p is a polynomial or constant-valued function and q is a polynomial function, is called a **rational function**.

EXERCISE 6.33

Determine which of the following are rational functions.

a) $f(x) = \dfrac{1}{x}$

b) $g(x) = \dfrac{x}{3}$

c) $h(x) = \dfrac{x-1}{4x}$

d) $j(x) = \dfrac{7}{x+2}$

e) $k(x) = \dfrac{\sqrt{x}}{1-2x}$

f) $m(x) = \dfrac{x^3 - x + 1}{x^2 + 1}$

SOLUTION 6.33

The functions f, h, j, and m are rational functions. The function g is not a rational function since its denominator is not a polynomial function; the denominator is a constant-valued function. The function k is not a rational function since its numerator is not a constant-valued or polynomial function; the numerator is a square root function.

From an examination of the definition of a rational function, we determine the following.

The Domain of a Rational Function:

The domain of a rational function is the set of all real numbers that are not *real* zeros of the polynomial in its denominator.

EXERCISE 6.34

Determine the domain of each of the following rational functions.

a) $f(x) = \dfrac{2x}{x+4}$

b) $g(x) = \dfrac{x^3 - 3x + 7}{x^2 - 1}$

c) $h(x) = \dfrac{5x - 77}{x^2 - 9}$

d) $j(x) = \dfrac{3}{3 - x}$

e) $k(x) = \dfrac{x^5}{x^6 + 5}$

SOLUTION 6.34

a) The only real zero of $x + 4$ is -4. Hence, the domain of f is $(-\infty, -4) \cup (-4, +\infty)$.

b) The real zeros of $x^2 - 1$ are 1 and -1. Hence, the domain of g is $(-\infty, -1) \cup (-1, 1) \cup (1, +\infty)$.

c) The real zeros of $x^2 - 9$ are 3 and -3. Hence, the domain of h is $(-\infty, -3) \cup (-3, 3) \cup (3, +\infty)$.

d) The only real zero of $3 - x$ is 3. Hence, the domain of j is $(-\infty, 3) \cup (3, +\infty)$.

e) There are no real zeros for $x^6 + 5$. Hence, the domain of k is $(-\infty, +\infty)$.

EXERCISE 6.35

Graph the function R defined by $R(x) = \dfrac{x^2 - 4}{x - 2}$.

SOLUTION 6.35

R is a rational function since it is the quotient of two polynomial functions. Further, the polynomial in the denominator has a real zero, 2. Hence, the domain of R is $(-\infty, 2) \cup (2, +\infty)$. There is no point on the graph of R whose x-coordinate is equal to 2. Also, the polynomial in the numerator is factorable as $(x - 2)(x + 2)$. Therefore,

$$R(x) = \frac{x^2 - 4}{x - 2} = \frac{(x - 2)(x + 2)}{x - 2} = x + 2$$

for all real values of x other than $x = 2$.

Now, consider the function L defined by $L(x) = x + 2$. L is a linear function whose graph is an oblique line with a slope of 1 and a y-intercept of 2. (See Figure 6.24a.) Therefore, the graph of R is the same oblique line as the graph of L, but with the point (2, 4) removed. (See Figure 6.24b.)

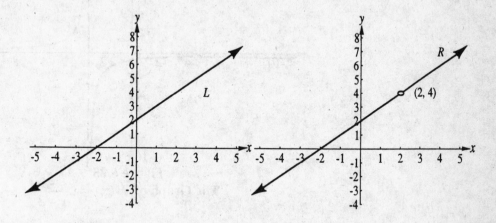

Figure 6.24

a) **The Graph of** $L(x) = x + 2$ b) **The Graph of** $R(x) = \dfrac{x^2 - 4}{x - 2}$

EXERCISE 6.36

Graph the function $H(x) = \dfrac{2}{x - 3}$.

SOLUTION 6.36

H is a rational function with domain $(-\infty, 3) \cup (3, +\infty)$. Hence, there is no point on the graph of H with x-coordinate equal to 3. However, this function, unlike the one in the previous exercise, does not simplify. If we select values of x close to 3, $H(x)$ is defined. If $x < 3$, then $H(x)$ is negative. If $x > 3$, then $H(x)$ is positive. Hence, the graph of H will be below the x-axis immediately to the left of the line with equation $x = 3$, and above the x-axis immediately to the right of that line. Moreover, the graph of H will not cross the line $x = 3$. (See Figure 6.25.) Such a line is called a **vertical asymptote**.

For all values of $x < 3$, $H(x)$ is negative and the graph of H will be below the x-axis. For all values of $x > 3$, $H(x)$ is positive and the graph will be above the x-axis. There are no values of x, then, for which $H(x)$ is equal to 0, and therefore, the graph of H will not cross the x-axis. (See Figure 6.25.) The x-axis (the line with equation $y = 0$) is called a **horizontal asymptote** for this graph.

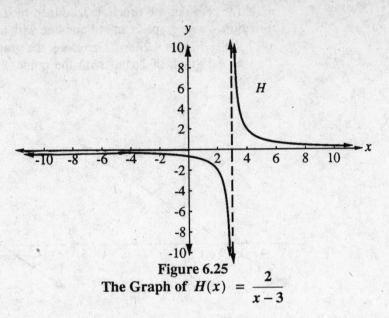

Figure 6.25
The Graph of $H(x) = \dfrac{2}{x-3}$

Zeros of a Rational Expression

When graphing a rational function, it is helpful to determine the zeros of the rational expression. If $R(x) = \dfrac{p(x)}{q(x)}$, then $R(x) = 0$ whenever $p(x) = 0$ and $q(x) \neq 0$. If $p(x) = 0$ and $q(x) = 0$, then $R(x)$ is not defined.

> **Definition:**
>
> A zero of a rational expression is a zero of its numerator that is not also a zero of its denominator.

EXERCISE 6.37

Determine all real zeros of $W(x) = \dfrac{(x-2)(x+3)(x+1)}{(x-2)(x-1)}$.

SOLUTION 6.37

Since the polynomials in the numerator and denominator are already factored, we determine that the real zeros of the numerator are –3, –1, and 2. The real zeros of the denominator are 1 and 2. Therefore, the real zeros of $W(x)$ are –3 and –1.

EXERCISE 6.38

Determine all real zeros of $T(x) = \dfrac{x^2 - 1}{x^2 - 5x + 6}$.

SOLUTION 6.38

The expression $T(x)$ can be written in factored form as

$$T(x) = \frac{(x-1)(x+1)}{(x-2)(x-3)}$$

The real zeros of the numerator are -1 and 1. The real zeros of the denominator are 2 and 3. Hence, the real zeros of $T(x)$ are -1 and 1.

EXERCISE 6.39

Determine all real zeros of $J(x) = \dfrac{x^2+1}{x^2-1}$.

SOLUTION 6.39

The numerator of $J(x)$ has no real zeros. Hence, there are no real zeros of $J(x)$.

When attempting to graph rational functions, every effort should be made to determine vertical and horizontal asymptotes to the graph of the function, if they exist. The following rules should be helpful.

Vertical Asymptotes

> **Rule for Determining Vertical Asymptotes:**
>
> Let R be a rational function defined by $R(x) = \dfrac{p(x)}{q(x)}$.
>
> If there exists a real number x_0 such that x_0 is a real zero of $q(x)$ with greater multiplicity than it has as a real zero of $p(x)$, then the line with equation $x = x_0$ is a vertical asymptote for the graph of R.

EXERCISE 6.40

Determine all vertical asymptotes, if any, for the graph of $R(x) = \dfrac{x+2}{x-3}$.

SOLUTION 6.40

The only real zero of the denominator of $R(x)$ is 3. Since 3 is not a real zero of the numerator of $R(x)$, then the line with equation $x = 3$ is a vertical asymptote for the graph of R.

EXERCISE 6.41

Determine all vertical asymptotes, if any, for the graph of

$$T(x) = \frac{x^3 + 3x^2 - x - 3}{x^2 - x - 6}.$$

SOLUTION 6.41

The numerator of $T(x)$ is factorable as $(x-1)(x+1)(x+3)$ and the denominator is factorable as $(x-3)(x+2)$. Hence, the real zeros of the denominator are -2 and 3. Neither of these is a zero of the numerator. Hence, the graph of T has two vertical asymptotes: the lines with equations $x = -2$ and $x = 3$.

EXERCISE 6.42

Determine all vertical asymptotes, if any, for the graph of the function V defined by $V(x) = \dfrac{x^2 - 4}{x^2 - 4x + 4}$.

SOLUTION 6.42

The numerator of $V(x)$ is factorable as $(x-2)(x+2)$. The denominator is factorable as $(x-2)(x-2)$. The only real zero of the denominator is 2, with a multiplicity two. The real zeros of the numerator are -2 and 2. Since 2 is a real zero of the denominator with a greater multiplicity than it is a zero of the numerator, then the graph of V has a vertical asymptote—the line with equation $x = 2$.

Horizontal Asymptotes

Rule for Determining Horizontal Asymptotes:

Let $p(x) = a_m x^m + a_{m-1} x^{m-1} + a_{m-2} x^{m-2} + \cdots + a_1 x + a_0$ be a polynomial of degree m, and let
$$q(x) = b_n x^n + b_{n-1} x^{n-1} + b_{n-2} x^{n-2} + \cdots + b_1 x + b_0$$
be a polynomial of degree n. Let $y = R(x) = \dfrac{p(x)}{q(x)}$. Then:

- If $m < n$, then the line $y = 0$ (i.e., the x-axis) is a horizontal asymptote for the graph of R.
- If $m = n$, then the line with equation $y = \dfrac{a_m}{b_n}$ is a horizontal asymptote for the graph of R.
- If $m > n$, then the graph of R does not have a horizontal asymptote.

EXERCISE 6.43

If the graph of the function $y = H(x) = \dfrac{x - 3}{x^2 + 1}$ has a horizontal asymptote, determine its equation.

SOLUTION 6.43

Since the degree of the numerator of H (which is 1) is less than the degree of the denominator (which is 2), the graph of H has a horizontal asymptote—the line with equation $y = 0$ (which is the x-axis).

EXERCISE 6.44

If the graph of the function $y = R(x) = \dfrac{5}{2x + 7}$ has a horizontal asymptote, determine its equation.

SOLUTION 6.44

Since the degree of the numerator of R (which is 0) is less than the degree of the denominator (which is 1), then the graph of R has a horizontal asymptote—the line with equation $y = 0$ (which is the x-axis).

EXERCISE 6.45

If the graph of the function $y = W(x) = \dfrac{2x + 9}{3x - 1}$ has a horizontal asymptote, determine its equation.

SOLUTION 6.45

Since the degree of the numerator of W (which is 1) is equal to the degree of the denominator (which is also 1), then the graph of W has a horizontal asymptote. We set y equal to the quotient of the leading coefficient of the numerator (which is 2) divided by the leading coefficient of the denominator (which is 3). Therefore, the horizontal asymptote is the line with equation $y = \dfrac{2}{3}$.

EXERCISE 6.46

If the graph of the function $y = J(x) = \dfrac{4x^5 - 5x + 7}{x^2 + 3}$ has a horizontal asymptote, determine its equation.

SOLUTION 6.46

Since the degree of the numerator (which is 5) is greater than the degree of the denominator (which is 2), then the graph of J does not have a horizontal asymptote.

Oblique Asymptotes

If the graph of a rational function does not have a horizontal asymptote, it may have an oblique asymptote.

Rule for Determining Oblique Asymptotes:

Let $y = R(x) = \dfrac{p(x)}{q(x)}$, where $p(x)$ and $q(x)$ are polynomials. If

i. the degree of $p(x)$ is *exactly one more than* the degree of $q(x)$,

ii. the division of $p(x)$ by $q(x)$ is *not exact*, and

iii. $\dfrac{p(x)}{q(x)} = Q(x) + \dfrac{r(x)}{q(x)}$ such that the degree of $r(x)$ is less than

the degree of $q(x)$,

then the line with equation $y = Q(x)$ is an oblique asymptote for the graph of R.

It should be noted that $Q(x)$ is a linear polynomial. We are dividing two polynomials, and the degree of the numerator is exactly one more than the degree of the denominator, resulting in a quotient which is a polynomial of the first degree. Hence, the asymptote, if it exists, is an oblique line.

EXERCISE 6.47

If the graph of $y = R(x) = \dfrac{x^3}{x^2+1}$ has an oblique asymptote, determine its equation.

SOLUTION 6.47

The degree of the numerator of $R(x)$ (which is 3) is exactly one more than the degree of the denominator (which is 2). Hence, the graph of R may have an oblique asymptote. Using long division, we divide x^3 by $x^2 + 1$, obtaining

$$\frac{x^3}{x^2+1} = x + \frac{-x}{x^2+1}.$$

The division is not exact. Hence, the line with equation $y = x$ is an oblique asymptote for the graph of R.

Graphs of Rational Functions

EXERCISE 6.48

Graph the rational function $y = R(x) = \dfrac{1}{x}$.

SOLUTION 6.48

R is the most basic of the rational functions. The domain of R is

$(-\infty, 0) \cup (0, +\infty)$. Further, 0 is a real zero of the denominator but not of the numerator. Hence, the line $x = 0$ (i.e., the *y*-axis) is a vertical asymptote for the graph of *R*. There are no real zeros of the numerator of $R(x)$; hence, there are no real zeros for $R(x)$. The degree of the numerator of $R(x)$ (which is 0) is less than the degree of its denominator (which is 1). Hence, the line with equation $y = 0$ (i.e., the *x*-axis) is a horizontal asymptote for the graph of *R*. There is no oblique asymptote. We also note that the function *R* is odd since

$$R(-x) = \frac{1}{-x} = \frac{-1}{x} = -R(x).$$

Hence, the graph of *R* is symmetric with respect to the origin. Finally, we note that if $x > 0$, then $R(x) > 0$ and the graph is completely above the *x*-axis. If $x < 0$, then $R(x) < 0$ and the graph is completely below the *x*-axis. Using a graphing calculator, we obtain the graph of *R*. (See Figure 6.26.)

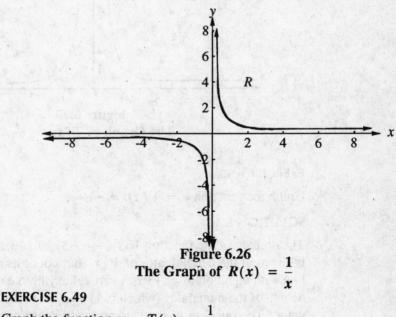

Figure 6.26
The Graph of $R(x) = \dfrac{1}{x}$

EXERCISE 6.49

Graph the function $y = T(x) = \dfrac{1}{x^2}$.

SOLUTION 6.49

The domain of *T* is $(-\infty, 0) \cup (0, +\infty)$. Further, 0 is a real zero of the denominator but not of the numerator. Hence, the line $x = 0$ (i.e., the *y*-axis) is a vertical asymptote for the graph of *T*. There are no real zeros of the numerator of $T(x)$; hence, there are no real zeros for $T(x)$. The degree of the numerator of $T(x)$ (which is 0) is less than the degree of its denominator (which is 2). Hence, the line with equation $y = 0$ (i.e., the *x*-axis) is a horizontal asymptote for the graph of *T*. There is no oblique asymptote. We also note that the function *T* is even since

$$T(-x) = \frac{1}{(-x)^2} = \frac{1}{x^2} = R(x)$$

Hence, the graph of *T* is symmetric with respect to the *y*-axis. Finally, we note that for all values of *x* in the domain of *T*, $T(x) > 0$. Hence, the graph of *T* is completely above the *x*-axis. Using a graphing calculator, we obtain the graph of *T*. (See Figure 6.27.)

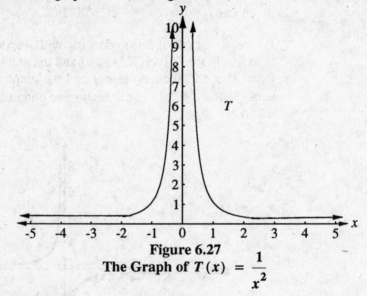

Figure 6.27
The Graph of $T(x) = \dfrac{1}{x^2}$

EXERCISE 6.50

Graph the function $y = V(x) = \dfrac{x-2}{x+3}$.

SOLUTION 6.50

The domain of the function *V* is $(-\infty, -3) \cup (-3, +\infty)$. We note that -3 is a zero of the denominator of $V(x)$ but not of its numerator. Hence, the line with equation $x = -3$ is a vertical asymptote of the graph of *V*. The degree of the numerator (which is 1) is equal to the degree of the denominator. The quotient of the leading coefficient in the numerator (which is 1) divided by the leading coefficient in the denominator (which is 1) is 1. Hence, the line with equation $y = 1$ is a horizontal asymptote for the graph of *V*. There is no oblique asymptote for its graph. We also note that 2 is a real zero of the numerator but not of the denominator; hence, 2 is a real zero of $V(x)$. The graph of *V* intersects the *x*-axis at the point $(2, 0)$. Using a graphing calculator, we obtain the graph of *V*. (See Figure 6.28.)

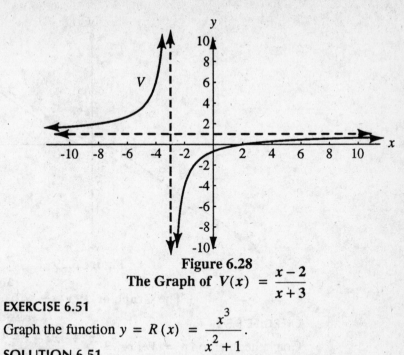

Figure 6.28
The Graph of $V(x) = \dfrac{x-2}{x+3}$

EXERCISE 6.51

Graph the function $y = R(x) = \dfrac{x^3}{x^2+1}$.

SOLUTION 6.51

Since $x^2 + 1 > 0$ for all real values of x, the domain of R is $(-\infty, +\infty)$. There are no real zeros for the denominator of $R(x)$; hence, there are no vertical asymptotes for the graph of R. The degree of the numerator of $R(x)$ (which is 3) is exactly one more than the degree of its denominator (which is 2). Hence, as noted in Exercise 6.47, the graph of R has an oblique asymptote—the line $y = x$. Further, we note that 0 is a real zero of the numerator of $R(x)$ but not of the denominator. Hence, the only real zero of $R(x)$ is 0, and the graph of R will cross the x-axis at the point $(0, 0)$. The function R is odd since

$$R(-x) = \frac{(-x)^3}{(-x)^2+1} = \frac{-x^3}{x^2+1} = -R(x)$$

and, hence, its graph is symmetric with respect to the origin. Using a graphing calculator, we obtain the graph of R. (See Figure 6.29.)

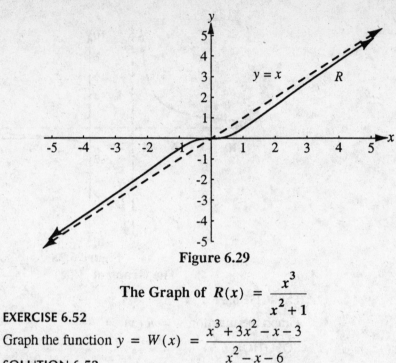

Figure 6.29

The Graph of $R(x) = \dfrac{x^3}{x^2 + 1}$

EXERCISE 6.52

Graph the function $y = W(x) = \dfrac{x^3 + 3x^2 - x - 3}{x^2 - x - 6}$

SOLUTION 6.52

From Exercise 6.41, we know that we can rewrite $W(x)$ as

$$y = W(x) = \frac{(x-1)\,(x+1)\,(x+3)}{(x-3)\,(x+2)}$$

The real zeros of the denominator are -2 and 3. Neither of these is a zero of the numerator. Hence, the graph of W has two vertical asymptotes: the lines with equations $x = -2$ and $x = 3$. The degree of the numerator (which is 3) is exactly one more than the degree of the denominator (which is 2). Dividing the numerator by the denominator, we have

$$y = W(x) = \frac{x^3 + 3x^2 - x - 3}{x^2 - x - 6} = (x+4) + \frac{3x + 21}{x^2 - x - 6}$$

and determine that the line with equation $y = x + 4$ is an oblique asymptote for the graph of W. Further, we determine that $W(x)$ has the real zeros -3, -1, and 1. Hence, the graph of W will intersect the x-axis at the points $(-3, 0)$, $(-1, 0)$, and $(1, 0)$. Using a graphing calculator, we obtain the graph of W. (See Figure 6.30.)

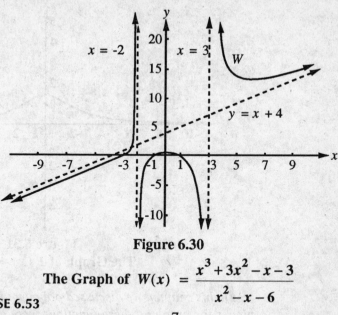

Figure 6.30

$$\text{The Graph of } W(x) = \frac{x^3 + 3x^2 - x - 3}{x^2 - x - 6}$$

EXERCISE 6.53

Graph the function $y = P(x) = \dfrac{7}{x^2 + 5}$.

SOLUTION 6.53

Since $x^2 + 5 > 0$ for all real values of x, there are no real zeros for the denominator of $P(x)$. Hence, the graph of P has no vertical asymptotes. Since the degree of the numerator (which is 0) is less than the degree of the denominator (which is 2), the graph of P has a horizontal asymptote— the line with equation $y = 0$ (i.e., the x-axis). Since both the numerator and the denominator of $P(x)$ are positive for all real values of x, $P(x)$ is also positive for all real values of x. The graph of P, then, is completely above the x-axis. When $x = 0$, $P(x) = \dfrac{7}{5}$. Hence, the graph of P will cross the y-axis at the point $\left(0, \dfrac{7}{5}\right)$. Further, note that P is an even function; hence, its graph is symmetric with respect to the y-axis. Using a graphing calculator, we obtain the graph of P. (See Figure 6.31.)

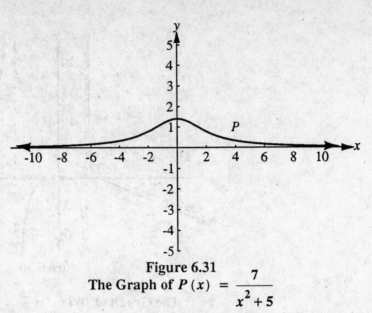

Figure 6.31
The Graph of $P(x) = \dfrac{7}{x^2 + 5}$

In this chapter, we discussed polynomial and rational functions. Techniques for graphing these functions were also discussed. These included synthetic division, the Remainder Theorem, the Factor Theorem, the Rational Zero Theorem, Descartes' Rule of Signs, the Intermediate Value Theorem, The Fundamental Theorem of Algebra, the Linear Factorization Theorem, and the Conjugate Complex Zero Theorem. Asymptotes for the graphs of rational functions were examined.

SUPPLEMENTAL EXERCISES

In Exercises 1–2, use synthetic division to determine the quotient and the remainder for each of the indicated divisions.

1. $(2x^4 - 3x^3 + 5x^2 - 6x + 1) \div (x - 1)$

2. $(2x^5 - 4x^3 + 2x + 2) \div (x + 1)$

In Exercises 3–4, use synthetic division to determine the following:

3. $A(3)$, if $A(x) = x^5 - 3x^4 + 5x + 1$

4. $B(-2)$, if $B(x) = x^4 - 2x^2 + 7$

In Exercises 5–6, use synthetic division to determine the following:

5. Is $(x - 3)$ a factor of $3x^5 - 2x^4 + 5x^2 + x - 11$?

6. Is -2 a zero of the polynomial $2x^3 - 2x^2 + 5x - 7$?

In Exercises 7–9, determine all rational zeros, if any, of the indicated polynomials.

7. $2x^3 - 4x^2 - 2x + 4$

8. $2x^3 - 8x^2 + 2x + 12$

9. $2x^4 - x^3 + 3x^2 - x - 5$

In Exercises 10–11, use Descartes' Rule of Signs to determine the number of positive and the number of negative zeros for the indicated polynomials.

10. $f(x) = 2x^5 - 3x - x^4 + 2x^3 + 7 - 3x^2$

11. $f(x) = 3x^2 - 12x^3 + 3x - 18x^4 + 108$

12. Determine the positive zero of the polynomial $x^3 - 2x - 6$, correct to the nearest tenth.

13. If the polynomial $x^5 - x^4 - 11x^3 + 12x^2$ has a negative zero, determine its value correct to the nearest tenth.

In Exercises 14–16, determine all of the real zeros, if any, for the given rational expressions.

14. $\dfrac{3x + 4}{5x - 1}$

15. $\dfrac{3x^2 - 9}{x + 8}$

16. $\dfrac{7}{2x + 5}$

For Exercises 17–20, determine all vertical asymptotes, all horizontal asymptotes, and all oblique asymptotes for the graphs of the following rational functions:

17. $y = R(x) = \dfrac{3x}{x^2 - 5x - 6}$

18. $y = Q(x) = \dfrac{2x - 3}{x^3 - 2x - x^2}$

19. $y = T(x) = \dfrac{9x^2}{x - 1}$

20. $y = V(x) = \dfrac{(2x - 1)(x + 3)}{(x + 4)(3x - 2)}$

ANSWERS TO SUPPLEMENTAL EXERCISES

1. Quotient $= 2x^3 - x^2 + 4x - 2$; remainder $= -1$
2. Quotient $= 2x^4 - 2x^3 - 2x^2 + 2x$; remainder $= 2$
3. 16
4. 15
5. No
6. No
7. -1, 1, and 2
8. -1, 2, and 3
9. There are none.
10. The number of positive zeros is four or two or none; the number of negative zeros is one.
11. The number of positive zeros is one; the number of negative zeros is three or one.
12. 2.2
13. -3.4
14. $\dfrac{-4}{3}$
15. $\pm\sqrt{3}$
16. There are none.
17. The vertical asymptotes for the graph of R are the lines with equations $x = 2$ and $x = 3$. The horizontal asymptote is the line with equation $y = 0$. There is no oblique asymptote.
18. The vertical asymptotes for the graph of Q are the lines with equations $x = -1$, $x = 0$, and $x = -1$. The horizontal asymptote is the line with equation $y = 0$. There is no oblique asymptote.
19. The vertical asymptote for the graph of T is the line with equation $x = 1$. There is no horizontal asymptote. The oblique asymptote is the line with equation $y = 9x + 9$.
20. The vertical asymptotes for the graph of V are the lines with equations $x = -4$ and $x = \dfrac{2}{3}$. The horizontal asymptote is the line with equation $y = \dfrac{2}{3}$. There is no oblique asymptote.

7

Exponential and Logarithmic Functions

So far in this text, we have discussed functions that can be classified as algebraic functions. Starting with this chapter, we shall examine non-algebraic functions, known as **transcendental functions**. Examples of such functions are the exponential, logarithmic, trigonometric, and inverse trigonometric functions.

In this chapter, we define exponential functions, as well as properties of exponents. We also introduce logarithmic functions as the inverses of exponential functions.

7.1 EXPONENTS

Exponents are the powers to which a base is raised. For instance, in the expressions 9^2, $4^{1/2}$, $(-5)^3$, $(ab)^{1/3}$, $(a+b)^{-3}$, and $3x^4$, the *bases* involved are 9, 4, −5, ab, $a+b$, and x. The *powers* of these bases are represented by the exponents 2, 1/2, 3, 1/3, −3, and 4, respectively. In the last expression, $3x^4$, 3 is the *numerical coefficient*. In each of the other expressions, the numerical coefficient is understood to be 1.

Positive Integer Exponents

Generally speaking, if a is a real number and n is a *positive* integer greater than one, then the expression a^n can be considered as the product of n factors, each equal to a. If $n = 1$, then $a^n = a^1 = a$.

Definition:

If a is a real number and n is a *positive* integer greater than 1, then

$$a^n = \underbrace{a \cdot a \cdot a \cdot \cdots \cdot a}_{n \text{ times}}.$$

If $n = 1$, then $a^n = a^1 = a$.

Properties of Exponents

The following properties follow immediately from the above definition. If a and b are real numbers and m and n are *positive* integers, then:

- $a^m a^n = a^{m+n}$
- $(a^m)^n = a^{mn}$
- $a^m \div a^n = a^{m-n}$, if $m > n$ and $a \neq 0$
- $a^m \div a^n = \dfrac{1}{a^{n-m}}$, if $m < n$ and $a \neq 0$
- $a^m \div a^n = 1$, if $m = n$ and $a \neq 0$
- $(ab)^n = a^n b^n$
- $\left(\dfrac{a}{b}\right)^n = \dfrac{a^n}{b^n}$, if $b \neq 0$

EXERCISE 7.1

Prove that if a is a real number and m and n are *positive* integers, then $a^m a^n = a^{m+n}$

SOLUTION 7.1

$$a^m a^n = \underbrace{(a \cdot a \cdot a \cdot \cdots \cdot a)}_{m \text{ times}} \underbrace{(a \cdot a \cdot a \cdot \cdots \cdot a)}_{n \text{ times}} \qquad \text{Definition}$$

$$= \underbrace{a \cdot a \cdot a \cdot \cdots \cdot a}_{(m+n) \text{ times}} \qquad \text{Rewrite}$$

$$= a^{m+n} \qquad \text{Definition}$$

EXERCISE 7.2

Prove that if a is a real number and m and n are *positive* integers, then $(a^m)^n = a^{mn}$.

SOLUTION 7.2

$$(a^m)^n = \underbrace{(a^m)\,(a^m)\,(a^m)\cdots(a^m)}_{n \text{ times}} \qquad \text{Definition}$$

$$= \underbrace{\underbrace{(a \cdot a \cdots a)}_{m \text{ times}}\ \underbrace{(a \cdot a \cdots a)}_{m \text{ times}}\ \cdots\ \underbrace{(a \cdot a \cdots a)}_{m \text{ times}}}_{n \text{ times}}$$

$$\text{Definition}$$

$$= \underbrace{a \cdot a \cdot a \cdots a}_{mn \text{ times}} \qquad \text{Rewrite}$$

$$= a^{mn} \qquad \text{Definition}$$

EXERCISE 7.3

Prove that if a is a *nonzero* real number and m and n are positive integers such that $m > n$, then $\dfrac{a^m}{a^n} = a^{m-n}$.

SOLUTION 7.3

$$\frac{a^m}{a^n} = \frac{\overbrace{a \cdot a \cdot a \cdots a}^{m \text{ times}}}{\underbrace{a \cdot a \cdot a \cdots a}_{n \text{ times}}} \qquad \text{Definition}$$

$$= \frac{\overbrace{(a \cdot a \cdot a \cdots a)}^{n \text{ times}}\ \overbrace{(a \cdot a \cdot a \cdots a)}^{(m-n) \text{ times}}}{\underbrace{a \cdot a \cdot a \cdots a}_{n \text{ times}}} \qquad \text{(Since } m > n.\text{)}$$

$$= \frac{\overbrace{(\cancel{a \cdot a \cdot a \cdots a})}^{n \text{ times}}\ \overbrace{(a \cdot a \cdot a \cdots a)}^{(m-n) \text{ times}}}{\underbrace{\cancel{a \cdot a \cdot a \cdots a}}_{n \text{ times}}} \qquad \text{(Since } a \neq 0.\text{)}$$

$$= (1) \underbrace{(a \cdot a \cdot a \cdot \cdots \cdot a)}_{(m-n) \text{ times}}$$

Common factor

$$= \underbrace{\frac{a \cdot a \cdot a \cdot \cdots \cdot a}{}}_{(m-n) \text{ times}}$$

Simplifying

$$= a^{m-n}$$

Definition

The other properties listed above are proven in a similar manner.

EXERCISE 7.4

Simplify each of the following by rewriting each as a single exponential expression.

a) $(2^2)(2^4)(2^5)$

b) $5^7 \div 5^3$

c) $7^3 \div 7^6$

d) $(3^2)^3$

e) $(3)^4 (9)^4$

f) $(3^2)(9^4)$

SOLUTION 7.4

a) $(2^2)(2^4)(2^5) = (2^{2+4})(2^5)$
$$= (2^6)(2^5)$$
$$= 2^{6+5}$$
$$= 2^{11}$$

b) $5^7 \div 5^3 = 5^{7-3} = 5^4$

c) $7^3 \div 7^6 = \dfrac{1}{7^{6-3}}$
$$= \dfrac{1}{7^3}$$

d) $(3^2)^3 = 3^{(2)(3)}$
$$= 3^6$$

e) $(3)^4 (9)^4 = [(3)(9)]^4$
$$= 27^4$$

f) $(3^2)(9^4) = (3^2)(3^2)^4$ (Rewriting $9 = 3^2$.)

$$= (3^2)(3^{(2)(4)})$$
$$= (3^2)(3^8)$$
$$= 3^{2+8}$$
$$= 3^{10}$$

Alternate solution:

$(3^2)(9^4) = (3^2)(3^2)^4$ (Rewriting $9 = 3^2$.)

$$= (3^2)^{1+4}$$
$$= (3^2)^5$$
$$= 3^{(2)(5)}$$
$$= 3^{10}$$

The Exponent 0

Consider the product $(a^0)(a^n)$, where a is a *nonzero* real number and n is a *positive* integer. We would like to maintain the properties of positive integer exponents. In particular, we would like to have

$$(a^0)(a^n) = a^{0+n} = a^n.$$

Since $a \neq 0$ and $n > 0$, then $a^n \neq 0$. Hence, dividing both sides of the preceding equation by a^n, we obtain

$$\frac{(a^0)(a^n)}{a^n} = \frac{a^n}{a^n} = 1.$$

Definition:

If a is a *nonzero* real number, then $a^0 = 1$.

EXERCISE 7.5

Simplify each of the following expressions.

a) 3^0

b) $(-2)^0$

c) -2^0

d) x^0

e) $(4x)^0$

f) $4x^0$

g) $(3xy)^0$

h) $(-5^0)^3$

SOLUTION 7.5

a) $3^0 = 1$

b) $(-2)^0 = 1$

c) $-2^0 = -(2^0) = -1$

d) $x^0 = 1$, if $x \neq 0$

e) $(4x)^0 = 1$, if $x \neq 0$

f) $4x^0 = 4(x^0) = 4(1) = 4$, if $x \neq 0$

g) $(3xy)^0 = 1$, if $x \neq 0$ **and** $y \neq 0$

h) $(-5^0)^3 = [-(5^0)]^3 = (-1)^3 = -1$

Negative Integer Exponents

When working with negative integer exponents, we would again like to preserve the properties of exponents already established. Consider, for example, the product of a^n and a^{-n}, where a is a *nonzero* real number and n is a *positive* integer. We would require that

$$(a^n)(a^{-n}) = a^{n+(-n)} = a^0 = 1.$$

Since $a^n \neq 0$, dividing both sides of the preceding equation by a^n, we obtain

$$\frac{(a^n)(a^{-n})}{a^n} = \frac{1}{a^n}$$

$$\frac{\cancel{(a^n)}(a^{-n})}{\underset{1}{\cancel{a^n}}} = \frac{1}{a^n}$$

or

$$a^{-n} = \frac{1}{a^n}$$

Definition:

If a is a *nonzero* real number and n is a *positive* integer, then

$$a^{-n} = \frac{1}{a^n}$$

We have now established that all of the properties of exponents introduced at the beginning of this section hold for positive, zero, and negative integer exponents.

EXERCISE 7.6

Simplify each of the following by (re)writing the expressions without negative exponents.

a) $3^3 \div 3^{-5}$

b) $(-3)^2 (-3)^{-5} (-3)^0$

c) $\dfrac{(4)^3 (-5)^{-4}}{(4)^{-2} (-5)^3}$

d) $(x^{-3})^2$

e) $(2x^{-1}yz^3)^{-2}$

SOLUTION 7.6

a) $3^3 \div 3^{-5} = 3^{3-(-5)}$

$\qquad = 3^8$

b) $(-3)^2 (-3)^{-5} (-3)^0 = (-3)^{2-5+0}$

$\qquad\qquad\qquad\qquad = (-3)^{-3}$

$\qquad\qquad\qquad\qquad = \dfrac{1}{(-3)^3}$

$\qquad\qquad\qquad\qquad = \dfrac{1}{-27}$

$\qquad\qquad\qquad\qquad = \dfrac{-1}{27}$

c) $\dfrac{(4)^3 (-5)^{-4}}{(4)^{-2} (-5)^3} = \dfrac{(4)^3}{(4)^{-2}} \cdot \dfrac{(-5)^{-4}}{(-5)^3}$

$\qquad\qquad\qquad = (4^{3-(-2)})(-5)^{-4-3}$

$\qquad\qquad\qquad = (4^5)(-5)^{-7}$

$\qquad\qquad\qquad = \dfrac{4^5}{(-5)^7}$

d) $(x^{-3})^2 = x^{(-3)(2)}$

$\qquad\qquad = x^{-6}$

$\qquad\qquad = \dfrac{1}{x^6}, \text{ if } x \neq 0$

e) $(2x^{-1}yz^3)^{-2} = (2)^{-2}(x^{-1})^{-2}(y)^{-2}(z^3)^{-2}$

$$= (2^{-2})(x^2)(y^{-2})(z^{-6})$$

$$= \frac{x^2}{(2^2)(y^2)(z^6)}$$

$$= \frac{x^2}{4y^2z^6}, \text{ if } x \neq 0, y \neq 0, \text{ and } z \neq 0$$

7.2 SCIENTIFIC NOTATION AND SIGNIFICANT DIGITS

When working with very large or very small numbers, it is convenient to use integer exponents and express the numbers in what is called **scientific notation**.

If N is a *positive* real number, we may write
$$N = p \times 10^k$$
such that $1 \leq p < 10$ and k is an integer. The value of p is determined by rewriting the number N with a decimal point after the first *nonzero* digit from the *left*. The integer k is determined by counting the number of digits the decimal point must be moved to obtain the original number, N. If the decimal point is moved to the right, then k will be positive; if it is moved to the left, then k will be negative.

EXERCISE 7.7

Rewrite each of the following numbers in scientific notation.

a) 126

b) 23,156

c) 0.000362

d) 0.019346

SOLUTION 7.7

a) $126 = 1.26 \times 10^2$

b) $23,156 = 2.3156 \times 10^4$

c) $0.000362 = 3.62 \times 10^{-4}$

d) $0.019346 = 1.9346 \times 10^{-2}$

Significant Digits

Observe that in rewriting a positive real number in scientific notation, we keep all of the basic digits of the given number. The digits required for writing the factor p in scientific notation are called the **significant digits**. For instance, 365, 0.00365, and 36,500 all have three significant digits since

$$365 = 3.65 \times 10^2,$$

$$0.00365 = 3.65 \times 10^{-3}, \text{ and}$$

$$36,500 = 3.65 \times 10^4.$$

If a number has n significant digits, we say that it is accurate to n places or that it has n-place accuracy.

Rounding

There are times when it is not necessary to maintain the number of significant digits given in a particular number. In such cases, the number may be **rounded** to fewer significant digits. For instance, the number 3.2437 has five significant digits, but can be expressed in terms of three significant digits, as 3.24, by dropping the digits 3 and 7. On the other hand, the number 3.2457, which also has five significant digits, when expressed in terms of three significant digits, has to be written as 3.25. The digits 5 and 7 are dropped, but the third significant digit is increased by one.

In general, to round a number to n significant digits, examine the digit immediately following the nth significant digit and proceed as follows:
- If the digit is less than 5, drop it and all digits following it.
- If the digit is greater than 5, drop it and all digits following it but increase the nth digit by one.
- If the digit is equal to 5 and
  a. not all remaining digits are equal to 0, drop it and all digits following it but increase the nth digit by one.
  b. all remaining digits are equal to 0, drop the digit and round the number up, if necessary, to make the last significant digit even.

EXERCISE 7.8

Round each of the following numbers, correct to three significant digits.
a) 2.3146
b) 47.286
c) 128.54
d) 12.350
e) 0.0079450

SOLUTION 7.8

a)   The third significant digit is 1. The digit immediately following it is 4, which is less than 5. Hence, 2.3146 is rounded to 2.31.

b)   The third significant digit is 2. The digit immediately following it is 8, which is greater than 5. Hence, 47.286 is rounded to 47.3.

c)   The third significant digit is 8. The digit immediately following it is 5. Not all remaining digits are equal to 0. Hence, 128.54 is rounded to 129.

d)   The third significant digit is 3. The digit immediately following it is 5. All remaining digits are 0. Hence, 12.350 is rounded to 12.4, which is even.

e)   The third significant digit is 4. The digit immediately following it is 5. All remaining digits are 0. Hence, 0.0079450 is rounded to 0.00794, which is even.

Approximate Numbers

Numbers that have been rounded to a specified number of significant digits are called **approximate numbers**. For instance, the exact number $\sqrt{3}$ can be written as 1.7, correct to two significant digits, or as 1.73, correct to three significant digits, and so forth. The symbol " $\approx$ " is used to denote "is approximately equal to." Therefore, $\sqrt{3} \approx 1.73$ is read "$\sqrt{3}$ is approximately equal to 1.73, correct to three significant digits."

When performing arithmetic operations with approximate numbers, the results obtained will be only as accurate as the least accurate of the numbers used.

7.3 ROOTS AND RADICALS

Before extending our discussion of exponents to rational exponents, we shall briefly discuss roots and radicals.

Definition:

Let a be a real number and n be a *positive* integer greater than 1. If there exists a real number b such that $a^n = b$, then a is called the **nth root** of b.

In particular, if $a^2 = b$, then a is called the **square root** of b. If $a^3 = b$, then a is called the **cube root** of b.

2 is a square root of 4 since $2^2 = 4$.

3 is a cube root of 27 since $3^3 = 27$.

-2 is a cube root of -8 since $(-2)^3 = -8$.

4^2 is a fifth root of 4^{10} since $(4^2)^5 = 4^{10}$.

When working with nth roots of real numbers, it is customary to introduce the symbol "$\sqrt[n]{b}$" where n is a positive integer greater than or equal to 2 and b is a real number. Such a symbol is called a **radical** and is read "the nth root of the real number b." The symbol "$\sqrt{}$" is called the **radical sign**; n is called the **index** of the radical; and b is called the **radicand**. When the square root is being considered, the index, 2, is generally omitted.

Every real number, except 0, has exactly n distinct nth roots. For instance, the two distinct square roots of 4 are 2 and -2 since $2^2 = (-2)^2 = 4$. Each of the square roots of 4 is real. However, the two distinct square roots of -4 are not real; $2i$ and $-2i$ are the square roots of -4.

In general, no even roots of negative real numbers are real. However, if n is a *positive odd integer*, then every real number has exactly one real number as an nth root. If n is a *positive even integer*, then every positive real number has exactly two real nth roots which are numerically equal but opposite in sign. If n is a *positive even integer*, then every negative real number has only imaginary nth roots.

For any computations involving roots of real numbers, we use only the principal nth roots.

Definition:

The principal nth root of a real number b is defined as follows:

- If b is *positive*, then the principal nth root of b is the positive nth root.
- If b is *negative* and n is *odd*, then the principal nth root of b is negative.
- If b is *negative* and n is *even*, then there is no principal nth root of b.

EXERCISE 7.9

Determine the indicated principal root, if it exists, of each of the following.

a) The principal square root of 25.

b) The principal cube root of -8.

c) The principal square root of -4.

SOLUTION 7.9

a) The two square roots of 25 are 5 and -5. Since 25 is positive, the principal square root of 25 is 5. Hence, $\sqrt{25} = 5$.

b) The number, -8, is negative and the root, 3, is odd. Hence, the principal cube root of -8 is -2. Hence, $\sqrt[3]{-8} = -2$. Also note that -2 is the only *real* cube root of -8; the other two are imaginary.

c) There is no principal square root of -4 since both of the square roots of -4 are imaginary.

EXERCISE 7.10

Determine all of the cube roots of -8.

SOLUTION 7.10

There are exactly three cube roots of -8; they are the solutions of the equation $x^3 = -8$, or equivalently, $x^3 + 8 = 0$. We know that -2 is the principal root of -8 and that, hence, -2 is a zero of the polynomial $x^3 + 8$. Therefore, $x + 2$ is a factor of the polynomial $x^3 + 8$. By synthetic division, we have

$$
\underline{-2|} \quad
\begin{array}{rrrr}
1 & 0 & 0 & 8 \\
  & -2 & 4 & -8 \\
\hline
1 & -2 & 4 & 0 = R
\end{array}
$$

Hence, $x^3 + 8 = (x + 2)(x^2 - 2x + 4)$. Setting $x^2 - 2x + 4 = 0$ and using the quadratic formula, we have

$$
x = \frac{2 \pm \sqrt{(-2)^2 - 4(1)(4)}}{2(1)}
$$

$$
= \frac{2 \pm \sqrt{4 - 16}}{2}
$$

$$
= \frac{2 \pm \sqrt{-12}}{2}
$$

$$
= \frac{2 \pm 2i\sqrt{3}}{2}
$$

$$
= 1 \pm i\sqrt{3}
$$

Hence, the three cube roots of -8 are -2, $1 + i\sqrt{3}$, and $1 - i\sqrt{3}$.

EXERCISE 7.11

Determine all fourth roots of 625.

SOLUTION 7.11

There are exactly four fourth roots of 625; they are the solutions of the equation $x^4 = 625$, or equivalently, $x^4 - 625 = 0$. We know that 5 is the principal fourth root of 625. Hence, 5 is a zero of $x^4 - 625$. Therefore, $x - 5$ is a factor of $x^4 - 625$. Using synthetic division, we have

$$
\begin{array}{r|rrrrr}
5) & 1 & 0 & 0 & 0 & -625 \\
   &   & 5 & 25 & 125 & 625 \\
\hline
   & 1 & 5 & 25 & 125 & 0 = R
\end{array}
$$

Since −5 is also a fourth root of 625, we have

$$
\begin{array}{r|rrrr}
-5) & 1 & 5 & 25 & 125 \\
    &   & -5 & 0 & 125 \\
\hline
    & 1 & 0 & 25 & 0 = R
\end{array}
$$

Hence, $x^4 - 25 = (x - 5)(x + 5)(x^2 + 25)$. Solving the equation $x^2 + 25 = 0$, we obtain $x = \pm 5i$. Hence, the four fourth roots of 625 are 5, −5, 5i, and −5i.

7.4 RATIONAL EXPONENTS

Exponents can also be rational numbers. If q is a *positive integer* greater than 1, then $1/q$ is a positive rational number. What meaning, if any, may be given to the symbol "$a^{1/q}$"? Again, we would like to preserve the properties established for positive integer exponents. In particular, we would like to have

$$
\underbrace{(a^{1/q})(a^{1/q})(a^{1/q})\cdots(a^{1/q})}_{q \text{ times}} = a^{\overbrace{1/q+1/q+1/q+\cdots+1/q}^{q \text{ times}}}
$$
$$
= a^{q/q}
$$
$$
= a^1
$$
$$
= a.
$$

The above is the equivalent of a being the qth root of $a^{1/q}$; that is $a^{1/q} = \sqrt[q]{a}$, or the principal qth root of a, if it exists.

Definition:
If a is a real number and q is a *positive integer* greater than 1, then $a^{1/q}$ is the **principal qth root** of a, if it exists.

The above can be extended to the rational exponent p/q, where p and q are both *positive integers* and p/q is simplified to lowest terms.

Definition:

If a is a real number and p and q are *positive integers* $(q > 1)$, then

$$a^{p/q} = \sqrt[q]{a^p} = (\sqrt[q]{a})^p.$$

The above can be extended to include negative rational exponents by using the properties similar to those for negative integer exponents.

EXERCISE 7.12

Evaluate each of the following.

a) $64^{1/3}$

b) $16^{3/4}$

c) $8^{-2/3}$

d) $(1/81)^{3/4}$

e) $(0.0064)^{3/2}$

f) $(-0.064)^{-2/3}$

SOLUTION 7.12

a) $64^{1/3} = \sqrt[3]{64} = 4$ since $4^3 = 64$.

b) $16^{3/4} = (\sqrt[4]{16})^3 = 2^3 = 8$

c) $8^{-2/3} = (8^{-2})^{1/3} = [(1/8)^2]^{1/3} = (1/64)^{1/3} = \sqrt[3]{1/64} = 1/4$

d) $(1/81)^{3/4} = (\sqrt[4]{1/81})^3 = (1/3)^3 = 1/27$

e) $(0.0064)^{3/2} = (\sqrt{0.0064})^3 = (0.08)^3 = 0.000512$

f) $(-0.064)^{-2/3} = (\sqrt[3]{-0.064})^{-2} = (-0.4)^{-2} = \dfrac{1}{(-0.4)^2} = \dfrac{1}{0.16}$

7.5 EXPONENTIAL FUNCTIONS

We shall now introduce the first of the transcendental functions encountered in this text. These are the Exponential functions.

Definition:

Let b be a *positive* number such that $b \neq 1$. The **Exponential function, base b**, denoted by Exp_b, is defined by $y = \text{Exp}_b(x) = b^x$ for all real values of x.

Some observations of the Exponential function, base b, are now in order.

- $\text{Exp}_b(x) = b^x$ is defined for all real values of x since, if $b > 0$, then b^x is a real number. Hence, the domain of Exp_b is the set of all real numbers. That is, the **domain** of Exp_b is $(-\infty, +\infty)$.

- If $x > 0$, then $b^x > 0$. If $x = 0$, then $b^x = 1$. If $x < 0$, then $b^x = \dfrac{1}{b^{-x}} > 0$. Hence, for all real values of x, $b^x > 0$. In fact, the range of Exp_b is the set of all positive real numbers. That is, the **range** of Exp_b is $(0, +\infty)$.

- The **graph** of Exp_b is found in the first and second quadrants only.

- If $0 < b < 1$, then b^x decreases as x increases; hence, for these values of b, Exp_b is a **decreasing** function.

- If $b > 1$, then b^x increases as x increases; hence, for these values of b, Exp_b is an **increasing** function.

- The **horizontal axis** is a **horizontal asymptote** for the graph of every Exp_b function.

- Every Exp_b function is **one-to-one**. Hence, every Exp_b function has an **inverse** as a function.

- The **graph** of every Exp_b function passes through the point with coordinates $(0, 1)$.

EXERCISE 7.13

Graph the Exponential function defined by $y = \text{Exp}_2(x) = 2^x$.

SOLUTION 7.13

Since $b = 2 > 1$, the graph of Exp_2 will be rising as we go through increasing values in its domain. In Table 7.1, we show some function values for the function. Using a graphing calculator, we determine the graph of the function. (See Figure 7.1.)

Table 7.1

x	-3	-2	-1	0	1	2	3
2^x	$\dfrac{1}{8}$	$\dfrac{1}{4}$	$\dfrac{1}{2}$	1	2	4	8

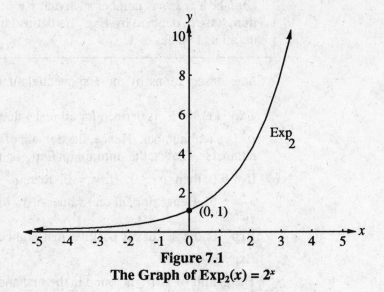

Figure 7.1
The Graph of $Exp_2(x) = 2^x$

EXERCISE 7.14

Graph the Exponential function defined by $y = Exp_{1/2}(x) = \left(\dfrac{1}{2}\right)^x$.

SOLUTION 7.14

Since $b = \dfrac{1}{2} < 1$, the graph of $Exp_{1/2}$ will be falling as we go through increasing values in its domain. In Table 7.2, we show some function values for the function. Using a graphing calculator, we determine the graph of the function. (See Figure 7.2.)

Table 7.2

x	-3	-2	-1	0	1	2	3
$\left(\dfrac{1}{2}\right)^x$	8	4	2	1	$\dfrac{1}{2}$	$\dfrac{1}{4}$	$\dfrac{1}{8}$

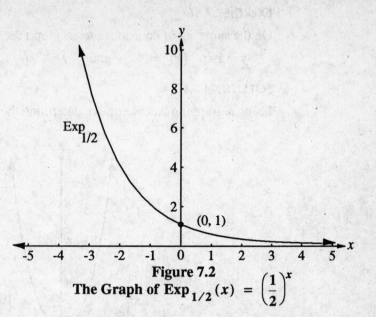

Figure 7.2
The Graph of $\text{Exp}_{1/2}(x) = \left(\dfrac{1}{2}\right)^x$

If you compare the graphs given in Figures 7.1 and 7.2, you will note that they are mirror images of each other with respect to the y-axis.

EXERCISE 7.15

On the same set of coordinate axes, graph the functions

$$y = \text{Exp}_3(x) = 3^x \quad \text{and} \quad y = \text{Exp}_{1/3}(x) = \left(\dfrac{1}{3}\right)^x.$$

SOLUTION 7.15

Using a graphing calculator, we determine the graphs. (See Figure 7.3.)

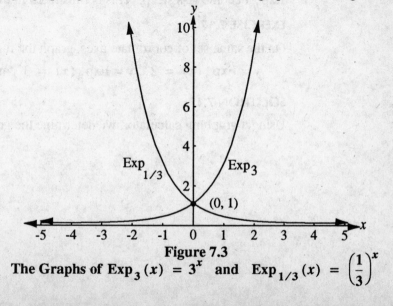

Figure 7.3
The Graphs of $\text{Exp}_3(x) = 3^x$ and $\text{Exp}_{1/3}(x) = \left(\dfrac{1}{3}\right)^x$

EXERCISE 7.16

On the same set of coordinate axes, graph the functions
$$y = \text{Exp}_4(x) = 4^x \quad \text{and} \quad y = \text{Exp}_{1/4}(x) = \left(\frac{1}{4}\right)^x.$$

SOLUTION 7.16

Using a graphing calculator, we determine the graphs. (See Figure 7.4.)

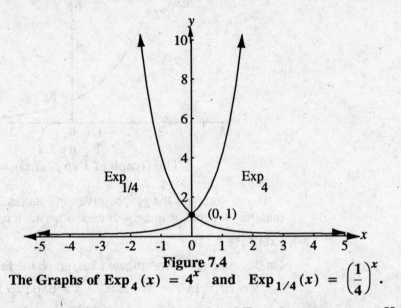

Figure 7.4
The Graphs of $\text{Exp}_4(x) = 4^x$ and $\text{Exp}_{1/4}(x) = \left(\frac{1}{4}\right)^x$.

Let $b > 1$. As b increases, the graphs of Exp_b become steeper. However, the opposite is true when $0 < b < 1$. As b increases, the graphs of Exp_b become less steep. This is illustrated in the the next two exercises.

EXERCISE 7.17

On the same set of coordinate axes, graph the functions
$$y = \text{Exp}_2(x) = 2^x, \; y = \text{Exp}_3(x) = 3^x, \text{ and } y = \text{Exp}_4(x) = 4^x.$$

SOLUTION 7.17

Using a graphing calculator, we determine the graphs. (See Figure 7.5.)

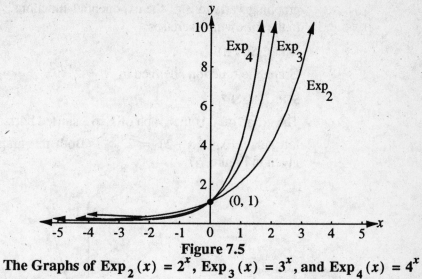

Figure 7.5

The Graphs of $\text{Exp}_2(x) = 2^x$, $\text{Exp}_3(x) = 3^x$, and $\text{Exp}_4(x) = 4^x$

EXERCISE 7.18

On the same set of coordinate axes, graph the functions $y = \text{Exp}_{1/2}(x) = \left(\frac{1}{2}\right)^x$, $y = \text{Exp}_{1/3}(x) = \left(\frac{1}{3}\right)^x$, and $y = \text{Exp}_{1/4}(x) = \left(\frac{1}{4}\right)^x$.

SOLUTION 7.18

Using a graphing calculator, we determine the graphs. (See Figure 7.6.)

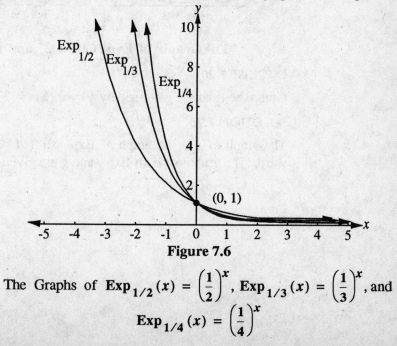

Figure 7.6

The Graphs of $\text{Exp}_{1/2}(x) = \left(\frac{1}{2}\right)^x$, $\text{Exp}_{1/3}(x) = \left(\frac{1}{3}\right)^x$, and

$$\text{Exp}_{1/4}(x) = \left(\frac{1}{4}\right)^x$$

The graphing techniques discussed in chapter 4 can also be used in graphing variations of the exponential functions. This will be illustrated in the following exercises.

EXERCISE 7.19

Graph the function f defined by $y = 2^{x+2}$.

SOLUTION 7.19

The graph of f is the graph of Exp_2 shifted horizontally two units to the left since $\text{Exp}_2(x+2) = 2^{x+2}$. Both the graphs of f and Exp_2 are given in Figure 7.7.

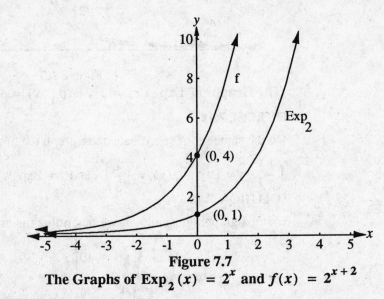

Figure 7.7
The Graphs of $\text{Exp}_2(x) = 2^x$ and $f(x) = 2^{x+2}$

EXERCISE 7.20

Graph the function g defined by $y = g(x) = 3^x - 2$.

SOLUTION 7.20

The graph of g is the graph of Exp_3 shifted vertically two units downward. The graphs of both Exp_3 and g are given in Figure 7.8.

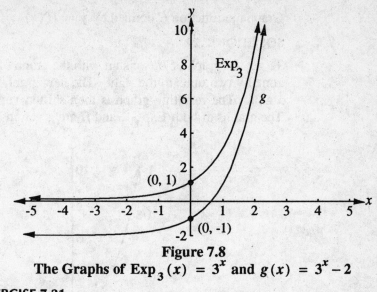

Figure 7.8

The Graphs of $\text{Exp}_3(x) = 3^x$ and $g(x) = 3^x - 2$

EXERCISE 7.21

Graph the function F defined by $y = F(x) = -\left(\dfrac{1}{2}\right)^x + 3$.

SOLUTION 7.21

The graph of F is the graph of $\text{Exp}_{1/2}$ reflected about the x-axis and then shifted vertically three units upward. The graphs of both $\text{Exp}_{1/2}$ and F are given in Figure 7.9.

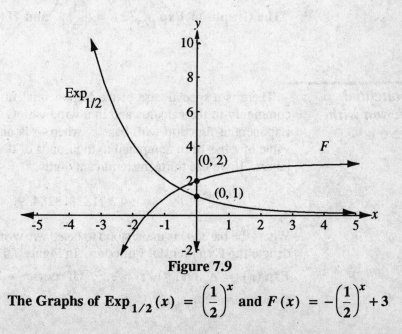

Figure 7.9

The Graphs of $\text{Exp}_{1/2}(x) = \left(\dfrac{1}{2}\right)^x$ and $F(x) = -\left(\dfrac{1}{2}\right)^x + 3$

EXERCISE 7.22

Graph the function H defined by $y = H(x) = -\left(\dfrac{1}{3}\right)^{x-2} + 4$.

SOLUTION 7.22

To get the graph of H, we start with the graph of $\text{Exp}_{1/3}$ and shift it horizontally two units to the right. The new graph is then reflected about the x-axis. The resulting graph is then shifted vertically four units upward. The graphs of both $\text{Exp}_{1/3}$ and H are given in Figure 7.10.

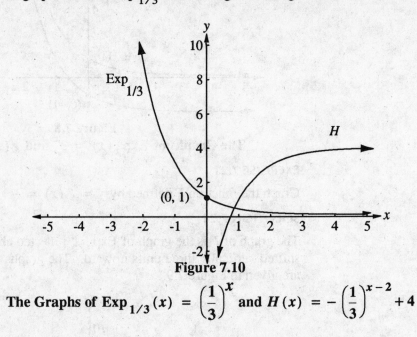

Figure 7.10

The Graphs of $\text{Exp}_{1/3}(x) = \left(\dfrac{1}{3}\right)^{x}$ and $H(x) = -\left(\dfrac{1}{3}\right)^{x-2} + 4$

Exponential Function with Base e

There is a special case of the Exponential functions that is used quite commonly in the calculus and in a wide variety of applications. It is the Exponential function with base e, where e is an irrational number. The value of e has been computed to thousands of decimal places, using computers. If we use thirteen significant digits,

$$e \approx 2.718281828459.$$

When the base, b, is understood to be e, we write Exp without a base to denote **the Exponential Function**. In Figure 7.11, we have the graph of $\text{Exp}(x) = e^{x}$ and $f(x) = e^{-x}$. Of course, e^{-x} is the same as $\left(\dfrac{1}{e}\right)^{x}$. Hence, $f = \text{Exp}_{1/e}$.

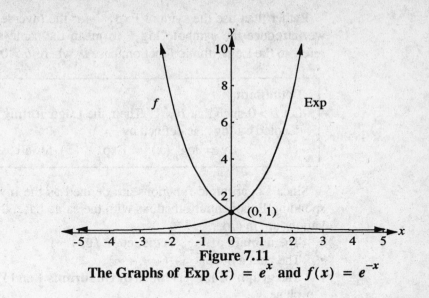

Figure 7.11

The Graphs of Exp $(x) = e^x$ and $f(x) = e^{-x}$

7.6 LOGARITHMIC FUNCTIONS

In the previous section, we observed that every Exponential function, base b, is one-to-one and has an inverse as a function. We also know that the graph of an inverse function can be obtained by reflecting the graph of the given function about the line $y = x$ in the xy-plane. The inverse function of the function Exp_b is denoted by Exp_b^{-1}. In Figure 7.12, we graph the functions $y = \text{Exp}_2(x) = 2^x$, $y = I(x) = x$, and $y = \text{Exp}_2^{-1}(x)$.

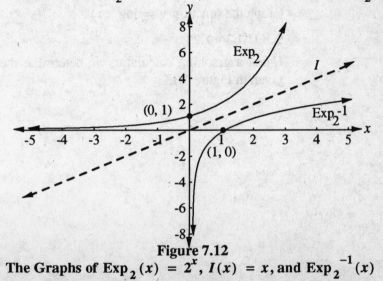

Figure 7.12

The Graphs of $\text{Exp}_2(x) = 2^x$, $I(x) = x$, and $\text{Exp}_2^{-1}(x)$

Rather than use the symbol Exp_b^{-1} for the inverse function of Exp_b, we introduce the symbol "$\log_b$" to mean the same as Exp_b^{-1}. $\log_b$ refers to the Logarithmic function, base b, where $b > 0$ and $b \neq 1$.

Definition:

Let $b > 0$ such that $b \neq 1$. Then, the **Logarithmic** function, base b, denoted by $\log_b$, is defined by
$$y = \log_b(x) = \text{Exp}_b^{-1}(x) \text{ for all } x > 0.$$

Since Logarithmic functions are defined as the inverses of the corresponding Exponential functions with the same base, the following observations are in order.

- The **domain** of $\log_b$ is the set of $(0, +\infty)$.
- The **range** of $\log_b$ is $(-\infty, +\infty)$.
- The **graph** of $\log_b$ is found in **Quadrants I** and **IV** in the coordinate plane.
- The **graph** of every $\log_b$ passes through the point with coordinates **(1, 0)**.
- The y-axis is a **vertical asymptote** for the graph of $\log_b$.
- Every $\log_b$ function is **one-to-one**.
- Every $\log_b$ function has an **inverse as a function**. Its inverse function is the corresponding Exponential function with the same base.
- Exp, without a base indicated, is **The Exponential Function**. The inverse function of Exp is the function $\log_e$, which is simply denoted by the symbol "**ln.**"

EXERCISE 7.23

Graph the function $y = \log_2(x)$.

SOLUTION 7.23

Using a graphing calculator, we determine the graph of $\log_2$, which is given in Figure 7.13.

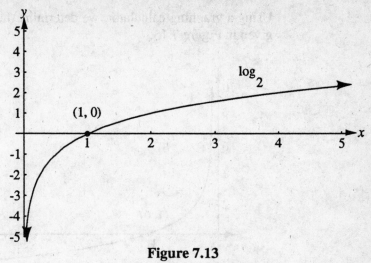

Figure 7.13
The Graph of $y = \log_2(x)$

EXERCISE 7.24

Graph the function $y = \log_7(x)$.

SOLUTION 7.24

Using a graphing calculator, we determine the graph of $\log_7$, which is given in Figure 7.14.

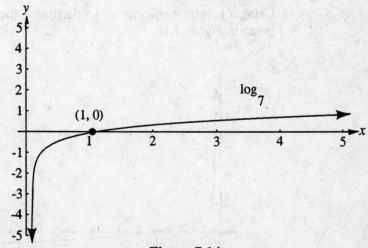

Figure 7.14
The Graph of $y = \log_7(x)$

EXERCISE 7.25

Graph the function $y = \log_{1/2}(x)$.

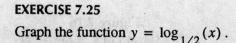

SOLUTION 7.25

Using a graphing calculator, we determine the graph of $\log_{1/2}$, which is given in Figure 7.15.

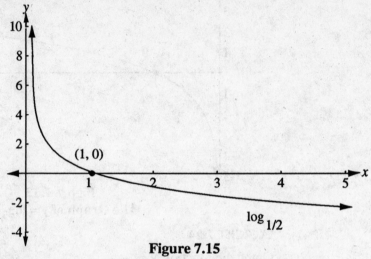

Figure 7.15
The Graph of $y = \log_{1/2}(x)$

EXERCISE 7.26

Graph the function $y = \log_{1/5}(x)$.

SOLUTION 7.26

Using a graphing calculator, we determine the graph of $\log_{1/5}$, which is given in Figure 7.16.

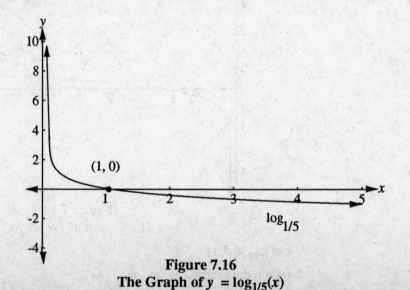

Figure 7.16
The Graph of $y = \log_{1/5}(x)$

In the remaining exercises in this section, we shall look at the graphs of variations of logarithmic functions.

EXERCISE 7.27

Graph the function defined by $y = \log_3(x + 4)$.

SOLUTION 7.27

The graph of the function $\log_3(x + 4)$ is the graph of the function $\log_3(x)$ shifted horizontally four units to the left. Both graphs are given in Figure 7.17.

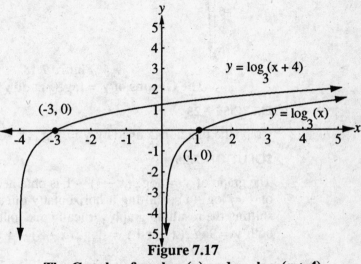

Figure 7.17

The Graphs of $y = \log_3(x)$ and $y = \log_3(x + 4)$

EXERCISE 7.28

Graph the function defined by $y = -\log_5(x) + 2$.

SOLUTION 7.28

To get the graph of $y = -\log_5(x) + 2$, we start with the graph of $y = \log_5(x)$ and reflect it about the x-axis. The resulting graph is then shifted vertically two units upward. The graphs of both $y = \log_5(x)$ and $y = -\log_5(x) + 2$ are given in Figure 7.18.

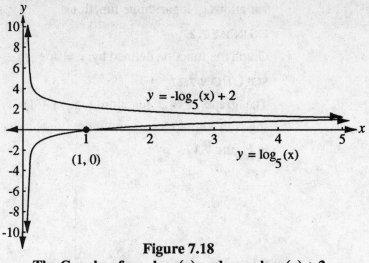

Figure 7.18
The Graphs of $y = \log_5(x)$ and $y = -\log_5(x) + 2$

EXERCISE 7.29

Graph the function defined by $y = \log_7(x-3) - 1$.

SOLUTION 7.29

The graph of $y = \log_7(x-3) - 1$ is obtained by starting with the graph of $y = \log_7(x)$, shifting it horizontally three units to the right and then shifting the resulting graph vertically one unit downward. The graphs of both $y = \log_7(x)$ and $y = \log_7(x-3) - 1$ are given in Figure 7.19.

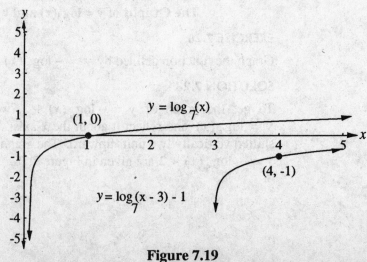

Figure 7.19
The Graphs of $y = \log_7(x)$ and $y = \log_7(x-3) - 1$

EXERCISE 7.30

Graph the function defined by $y = -\ln(x+1) + 3$.

SOLUTION 7.30

To obtain the graph of $y = -\ln(x+1) + 3$, we start with the graph of $y = \ln(x)$, shift it horizontally one unit to the left, and then reflect the resulting graph about the x-axis. The new graph is then shifted vertically three units upward. The graphs of both $y = \ln(x)$ and $y = -\ln(x+1) + 3$ are given in Figure 7.20.

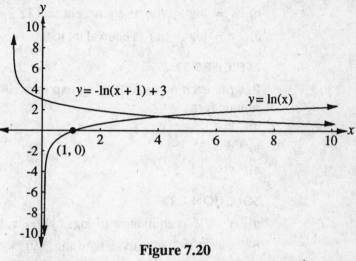

Figure 7.20
The Graphs of $y = \ln(x)$ and $y = -\ln(x+1) + 3$

7.7 PROPERTIES OF LOGARITHMS

In the previous section, we noted that the Logarithmic function, base b, is the inverse of the corresponding Exponential function, base b.

Logarithms

> **Definition:**
>
> Let x be a *positive* number. Let b be a *positive* number not equal to 1. Then, the **logarithm of x, base b**, denoted by $\log_b(x)$, is the exponent y to which the base b must be raised to produce the number x. Hence, $y = \log_b(x)$ is equivalent to $b^y = x$.

EXERCISE 7.31

Rewrite each of the following logarithmic equations in equivalent exponential form. (Assume that $x > 0$.)

a) $y = \log_7(x)$

b) $y = \log_{1/3}(x)$

c) $y = \log_{0.4}(x)$

SOLUTION 7.31

a) $y = \log_7(x)$ is equivalent to $7^y = x$.

b) $y = \log_{1/3}(x)$ is equivalent to $(1/3)^y = x$.

c) $y = \log_{0.4}(x)$ is equivalent to $(0.4)^y = x$.

EXERCISE 7.32

Rewrite each of the following exponential equations in equivalent logarithmic form.

a) $y = 3^x$

b) $y = 2^{x-2}$

c) $y = (1/5)^{3x}$

SOLUTION 7.32

a) $y = 3^x$ is equivalent to $\log_3(y) = x$, if $y > 0$.

b) $y = 2^{x-2}$ is equivalent to $\log_2(y) = (x-2)$, if $y > 0$.

c) $y = (1/5)^{3x}$ is equivalent to $\log_{1/5}(y) = 3x$, if $y > 0$.

Computations

EXERCISE 7.33

Compute $\log_2(8)$.

SOLUTION 7.33

Let $y = \log_2(8)$. Then, $2^y = 8$ which is equivalent to $2^y = 2^3$. Since the function Exp is one-to-one, we deduce that $y = 3$. Therefore, $\log_2(8) = 3$.

EXERCISE 7.34

Compute $\log_{1/2}(16)$.

SOLUTION 7.34

Let $y = \log_{1/2}(16)$. Then, $(1/2)^y = 16$, which is equivalent to $2^{-y} = 16$, which, in turn, is equivalent to $2^{-y} = 2^4$. We deduce that $-y = 4$, or $y = -4$. Therefore, $\log_{1/2}(16) = -4$.

EXERCISE 7.35

Compute:

a) $\ln(e)$

b) $\log_b(b)$

SOLUTION 7.35

a)   Let $y = \ln(e)$. Since $\ln(e)$ is equivalent to $\log_e(e)$, we have that $y = \log_e(e)$, which, in turn, is equivalent to $e^y = e$. But, $e^y = e$ is equivalent to $e^y = e^1$ from which we deduce that $y = 1$. Hence, $\ln(e) = 1$.

b)   Let $y = \log_b(b)$ which is equivalent to $b^y = b$, or $b^y = b^1$. We deduce that $y = 1$. Hence, $\log_b(b) = 1$.

From Exercise 7.35, we obtain the following:

Fact:	$\log_b(b) = 1$

EXERCISE 7.36

Compute $\log_b(1)$.

SOLUTION 7.36

Let $y = \log_b(1)$, which is equivalent to $b^y = 1$. But, $b^y = 1$ can be rewritten in equivalent form as $b^y = b^0$ since $b \neq 0$. We deduce that $y = 0$. Therefore, $\log_b(1) = 0$.

From Exercise 7.36, we obtain the following:

Fact:	$\log_b(1) = 0$

EXERCISE 7.37

Solve the exponential equation $4^x = 256$.

SOLUTION 7.37

$4^x = 256$

$4^x = 4^4$ (Since $4^4 = 256$.)

Therefore, $x = 4$.

EXERCISE 7.38

Solve the exponential equation $3^{x-5} = 27$.

SOLUTION 7.38

$$3^{x-5} = 27$$

$$3^{x-5} = 3^3$$

We deduce that $x - 5 = 3$.

$$x = 8$$

Therefore, $x = 8$.

EXERCISE 7.39

Solve the exponential equation $2^{3x-1} = \dfrac{1}{4}$.

SOLUTION 7.39

$$2^{3x-1} = \frac{1}{4}$$

$$2^{3x-1} = \left(\frac{1}{2}\right)^2$$

$$2^{3x-1} = 2^{-2}$$

We deduce that $3x - 1 = -2$.

$$3x = -1$$

$$x = \frac{-1}{3}$$

Therefore, $x = \dfrac{-1}{3}$.

EXERCISE 7.40

Solve the exponential equation $9^{2x} = 27^{3x+4}$.

SOLUTION 7.40

$$9^{2x} = 27^{3x+4}$$

$$\left(3^2\right)^{2x} = \left(3^3\right)^{3x+4}$$

$$3^{4x} = 3^{9x+12}$$

We deduce that $4x = 9x + 12$.

$$-5x = 12$$

$$x = \frac{-12}{5}$$

Therefore, $x = \dfrac{-12}{5}$.

Properties of Logarithms

We shall now discuss properties of logarithms. Let x and y be *positive* numbers. Let b be a *positive* number not equal to 1. Let k be a real number.

- $\log_b(xy) = \log_b(x) + \log_b(y)$

  (In words, the logarithm of a **product** of two *positive* numbers is equal to the **sum** of the logarithms of the factors.)

- $\log_b\left(\dfrac{x}{y}\right) = \log_b(x) - \log_b(y)$

  (In words, the logarithm of a **quotient** of two *positive* numbers is equal to the **difference** of the logarithms of the numbers.)

- $\log_b(x^k) = k \cdot \log_b(x)$

  (In words, the logarithm of a **power** of a *positive* number is equal to the **product** of the exponent in the power and the logarithm of the number.)

EXERCISE 7.41

Prove that if x and y are two *positive* numbers and b is a *positive* number not equal to 1, then $\log_b(xy) = \log_b(x) + \log_b(y)$.

SOLUTION 7.41

Let $u = \log_b(x)$ and $v = \log_b(y)$. Then, $b^u = x$ and $b^v = y$. By using properties of exponents, we have
$$xy = (b^u)(b^v)$$
$$xy = b^{u+v}$$
But, $xy = b^{u+v}$ is equivalent to $\log_b(xy) = u+v = \log_b(x) + \log_b(y)$. Hence, $\log_b(xy) = \log_b(x) + \log_b(y)$.

EXERCISE 7.42

Prove that if x and y are two *positive* numbers and b is a *positive* number not equal to 1, then $\log_b\left(\dfrac{x}{y}\right) = \log_b(x) - \log_b(y)$.

SOLUTION 7.42

Let $p = \log_b(x)$ and $q = \log_b(y)$. Then, $b^p = x$ and $b^q = y$. By using properties of exponents, we have $x/y = b^p/b^q = b^{p-q}$. But, $x/y = b^{p-q}$ is equivalent to $\log_b(x/y) = p-q$. Therefore, $\log_b(x/y) = \log_b(x) - \log_b(y)$.

EXERCISE 7.43

Let x be a *positive* number. Let b be a *positive* number not equal to 1. Let k be a real number. Prove that $\log_b(x^k) = k \cdot \log_b(x)$.

SOLUTION 7.43

Let $r = \log_b(x^k)$. Then,

$$b^r = x^k$$

$$(b^r)^{1/k} = (x^k)^{1/k} \qquad \text{(Raising each side to the } 1/k \text{ power.)}$$

$$b^{r/k} = x,$$

which is equivalent to

$$\log_b(x) = \frac{r}{k}$$

or

$$k \cdot \log_b(x) = r.$$

Substituting for r, we have

$$\log_b(x^k) = k \cdot \log_b(x)$$

EXERCISE 7.44

Express $\log_7(72)$ in terms of $\log_7(2)$ and $\log_7(3)$.

SOLUTION 7.44

$$\begin{aligned}
\log_7(72) &= \log_7[(8)(9)] \\
&= \log_7(8) + \log_7(9) \\
&= \log_7(2^3) + \log_7(3^2) \\
&= 3 \cdot \log_7(2) + 2 \cdot \log_7(3)
\end{aligned}$$

Therefore, $\log_7(72) = 3 \cdot \log_7(2) + 2 \cdot \log_7(3)$.

EXERCISE 7.45

Express $\frac{1}{3} \cdot \log_5(2) + 2 \cdot \log_5(4) - \log_5(8)$ as a single logarithm.

SOLUTION 7.45

$$\frac{1}{3} \cdot \log_5(2) + 2 \cdot \log_5(4) - \log_5(8)$$

$$= \log_5(2^{1/3}) + \log_5(4^2) - \log_5(8)$$

$$= \log_5[(2^{1/3})(4^2)] - \log_5(8)$$

$$= \log_5 \left[\frac{(2^{1/3})\,(4^2)}{(8)} \right]$$

$$= \log_5 \left[(2^{1/3})\,(2) \right]$$

$$= \log_5 (2^{4/3})$$

$$= \frac{4}{3} \cdot \log_5 (2)$$

EXERCISE 7.46

Express $\ln (10) + 2 \cdot \ln (\pi) - 3 \cdot \ln (e)$ as a single logarithm.

SOLUTION 7.46

$$\ln (10) + 2 \cdot \ln (\pi) - 3 \cdot \ln (e) = \ln (10) + \ln (\pi^2) - \ln (e^3)$$

$$= \ln (10\pi^2) - \ln (e^3)$$

$$= \ln \left(\frac{10\pi^2}{e^3} \right).$$

EXERCISE 7.47

Determine the value of $\dfrac{\log_3 (\sqrt{1/81}) - \log_5 (625)^{-2}}{\log_5 (25) \log_4 (\sqrt[3]{16})}$.

SOLUTION 7.47

Let $\log_3 (\sqrt{1/81}) = x$. Then,

$$3^x = \sqrt{1/81} = 1/9 = 9^{-1} = (3^2)^{-1} = 3^{-2}.$$

Therefore, $x = -2$.

Let $\log_5 (625)^{-2} = y$. Then,

$$5^y = 625^{-2} = (5^4)^{-2} = 5^{-8}.$$

Therefore, $y = -8$.

Let $\log_5 (25) = u$. Then,

$$5^u = 25 = 5^2.$$

Therefore, $u = 2$.

Let $\log_4 (\sqrt[3]{16}) = v$. Then,

$$4^v = \sqrt[3]{16} = 16^{1/3} = (4^2)^{1/3} = 4^{2/3}.$$

Therefore, $v = 2/3$.

Substituting the computed values for x, y, u, and v, we now have

$$\frac{\log_3(\sqrt{1/81}) - \log_5(625)^{-2}}{\log_5(25)\log_4(\sqrt[3]{16})} = \frac{-2 - (-8)}{2(2/3)}$$

$$= \frac{6}{4/3}$$

$$= \frac{9}{2}.$$

Change of Base

Computing logarithms to various bases requires the use of logarithmic tables or a calculator/computer. Generally, you can find tables for base 10 and base e in mathematical handbooks. However, even if only a single base table existed, you would be able to represent the logarithm of any positive number to any base in terms of the given base. We will now show this to be true.

Change of Base:
If a, b, and x are *positive* numbers such that $a \neq 1$ and $b \neq 1$, then
$$\log_a(x) = \frac{\log_b(x)}{\log_b(a)}$$

To prove the above statement, let $p = \log_a(x)$. Then,

$$a^p = x.$$

Therefore,

$$\log_b(a^p) = \log_b(x) \qquad \text{(One-to-one property of logarithms)}$$

and

$$p \cdot \log_b(a) = \log_b(x).$$

Hence,

$$p = \frac{\log_b(x)}{\log_b(a)}$$

or

$$\log_a(x) = \frac{\log_b(x)}{\log_b(a)}.$$

EXERCISE 7.48

Rewrite each of the given expressions in equivalent form using base 7 logarithms.

a) $\log_3(4)$

b) $\log_2(5)$

c) $\log_8(23)$

SOLUTION 7.48

a) $\log_3(4) = \dfrac{\log_7(4)}{\log_7(3)}$

b) $\log_2(5) = \dfrac{\log_7(5)}{\log_7(2)}$

c) $\log_8(23) = \dfrac{\log_7(23)}{\log_7(8)}$

*I*n *this chapter, we discussed exponential and logarithmic functions and the relationship between them. Properties of exponents and properties of logarithms were examined. Scientific notation and rounding were also discussed.*

SUPPLEMENTAL EXERCISES

For Exercises 1–4, indicate whether the given function is an increasing function, decreasing function, or neither.

1. $y = A(x) = 2^{x-3}$

2. $y = B(x) = \log_{1/3}(2x+1)$

3. $y = C(x) = -3^{-x}$

4. $y = D(x) = -2 \cdot \log_7(1-3x)$

For Exercises 5–8, determine the range for each of the indicated functions.

5. $y = E(x) = 4^{x+1} - 3$

6. $y = F(x) = -\log_4(x-5) + 7$

7. $y = G(x) = -5^{-x} - 5$

8. $y = H(x) = \ln(3x-7) + \pi$

For Exercises 9–11, evaluate the given expression.

9. $\log_2(64)$

10. $\ln(e^3)$

11. $\log_{1/3}(243)$

For Exercises 12–15, solve each of the given equations for x.

12. $5^{2x+1} = 625$

13. $3^{2-4x} = \dfrac{1}{3}$

14. $4^{1-3x} = 16^{x+2}$

15. $49^{1-2x} = 343^{4x}$

For Exercises 16–20, determine the inverse function for each of the functions given.

16. $y = \log_9(x)$

17. $y = e^x$

18. $y = \text{Exp}_{0.2}(x)$

19. $y = \text{Exp}_{0.9}^{-1}(x)$

20. $y = \log_4^{-1}(x)$

ANSWERS TO SUPPLEMENTAL EXERCISES

1. Increasing

2. Decreasing

3. Increasing

4. Decreasing

5. $(-3, +\infty)$

6. $(-\infty, +\infty)$

7. $(-\infty, -5)$

8. $(-\infty, +\infty)$

9. 6

10. 3

11. –5

12. $\dfrac{3}{2}$

13. $\dfrac{3}{4}$

14. $\dfrac{-3}{5}$

15. $\dfrac{1}{8}$

16. $y = 9^x$

17. $y = \ln(x)$

18. $y = \log_{0.2}(x)$

19. $y = 0.9^x$

20. $y = \log_4(x)$

8

Circular Functions

Historically, the study of trigonometry evolved as it related to the measurement of angles and the ratio of the measure of sides of triangles. However, in this text, we introduce trigonometric functions as circular functions associated with the Wrapping function.

8.1 THE WRAPPING FUNCTION

> **Definition:**
> A **unit circle** is a circle with its center at the origin of a coordinate system and having a radius of 1.

Each point on the circumference of a unit circle is one unit from the origin. Therefore, by the Pythagorean Theorem, the coordinates, (x, y), of any point on the circumference of the unit circle, in the xy-plane, must satisfy the equation $x^2 + y^2 = 1$. (See Figure 8.1.)

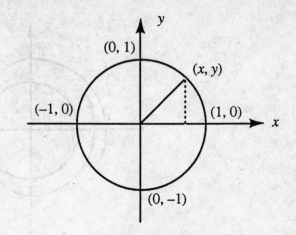

Figure 8.1
The Graph of the Circle with Equation $x^2 + y^2 = 1$

Suppose that we were to "wrap" the real number line around the unit circle as follows:

1. Let 0 on the number line coincide with the point $(1, 0)$ on the circle.
2. Let the *positive* numbers on the number line coincide with points on the unit circle as we proceed in a *counterclockwise* (or positive) direction along the circumference from the point $(1, 0)$.
3. Let the *negative* numbers on the number line coincide with points on the unit circle as we proceed in a *clockwise* (or negative) direction from the point $(1, 0)$.
4. The number line and the coordinate axes of the plane would all have the same scale. (See Figure 8.2.)

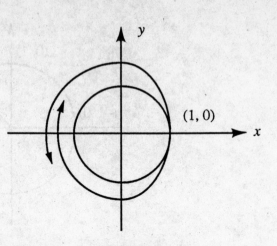

Figure 8.2
"Wrapping" the Number Line About the Unit Circle

Note that each point on the number line is associated with a unique point on the unit circle. However, each point on the unit circle is associated with infinitely many points on the number line. The procedure of "wrapping" the number line about the unit circle defines a function which is called the Wrapping function.

Definition:
Let C denote the set of points on the unit circle, the center of which is at the origin in the xy-plane. The **Wrapping function**, denoted by W, is defined as the mapping of the real numbers (on the number line) into the set C, in the manner described above.

We now make the following observations about the Wrapping function, W.

- The **domain** of W is the set $(-\infty, +\infty)$.
- The **range** of W is the set of *all* ordered pairs (x, y) such that $x^2 + y^2 = 1$.
- The function W is **not one-to-one**. For instance, the element $(1, 0)$ in the range of W is paired with both 0 and 2π in the domain of W.
- The function W does **not** have an **inverse as a function** since W is not one-to-one.

Function Values

Since the circumference of the unit circle is equal to $2\pi r = 2\pi(1) = 2\pi$, it is fairly easy to determine select function values for W. For instance, since the function W maps both 0 and 2π onto the point with coordinates $(1, 0)$, we have

$$W(0) = (1, 0) \quad \text{and} \quad W(2\pi) = (1, 0)$$

Since $W(\pi)$ is halfway around the unit circle from $(1, 0)$, we have

$$W(\pi) = (-1, 0)$$

Similarly,

$$W(\pi/2) = (0, 1) \quad \text{and} \quad W(-\pi/2) = (0, -1)$$

$$W(3\pi/2) = (0, -1) \quad \text{and} \quad W(-3\pi/2) = (0, 1)$$

Examining the function W further, we determine that if n is an *integer*, then

$$W(0) = W(2\pi) = W(4\pi) = W(6\pi) = \cdots = W(2n\pi) = (1, 0)$$

and

$$W(0) = W(-2\pi) = W(-4\pi) = W(-6\pi) = \cdots = W(-2n\pi) = (1, 0)$$

indicating that all real numbers that are integer multiples of 2π get mapped onto the point $(1, 0)$ under the function W.

Similarly, all real numbers that are $(4n + 1)$-multiples of $\pi/2$, where n is an integer, get mapped onto the point $(0, 1)$ under the function W. That is,

$$W(\pi/2) = W(5\pi/2) = W(9\pi/2) = W(-3\pi/2) = W(-7\pi/2)$$

$$= (0, 1) .$$

EXERCISE 8.1

Determine the function values for each of the following:

a) $W(3\pi)$

b) $W(5\pi/2)$

c) $W(-7\pi/2)$

d) $W(23\pi/2)$

e) $W(-37\pi/2)$

f) $W(52\pi)$

SOLUTION 8.1

a) $W(3\pi) = (-1, 0)$

b) $W(5\pi/2) = (0, 1)$

c) $W(-7\pi/2) = (0, 1)$

d) $W(23\pi/2) = (0, -1)$

e) $W(-37\pi/2) = (0, -1)$

f) $W(52\pi) = (1, 0)$

Now, let us try to determine the value of $W(\pi/4)$. Since $\pi/4$ is one-half of the distance from 0 to $\pi/2$ along the circumference of the unit circle, the point $W(\pi/4)$ is the midpoint of the arc joining the points $W(0)$ and $W(\pi/2)$. Consequently, $W(\pi/4)$ lies on the line with the equation $y = x$. (See Figure 8.3.) Therefore, the coordinates of the point $W(\pi/4)$ must satisfy the equation $x^2 + y^2 = 1$ with $y = x$. Hence,

$$x^2 + y^2 = 1$$

becomes

$$x^2 + x^2 = 1$$

or

$$2x^2 = 1$$

or

$$x^2 = \frac{1}{2}.$$

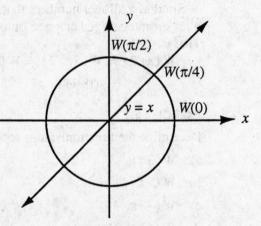

Figure 8.3
$W(\pi/4)$ **Lies on the Line** $y = x$

Solving this last equation for x, and observing that x must be positive, we obtain

$$x = \frac{\sqrt{2}}{2}$$

We now have

$$W(\pi/4) = \left(\frac{\sqrt{2}}{2}, \frac{\sqrt{2}}{2}\right).$$

EXERCISE 8.2

Determine the coordinates of $W(3\pi/4)$, $W(-\pi/4)$, and $W(5\pi/4)$.

SOLUTION 8.2

Referring to Figure 8.4, we observe that $W(3\pi/4)$ is the midpoint of the arc joining the points $W(\pi/2)$ and $W(\pi)$. Therefore, $W(3\pi/4)$ has the same y value $-\dfrac{\sqrt{2}}{2}-$ as $W(\pi/4)$. Since $W(\pi/4)$ and $W(3\pi/4)$ are symmetric with respect to the y-axis, and since the x value of $W(\pi/4)$ is $\sqrt{2}/2$, the x value of $W(3\pi/4)$ is $-\sqrt{2}/2$. Hence,

$$W(3\pi/4) = \left(\frac{-\sqrt{2}}{2}, \frac{\sqrt{2}}{2}\right).$$

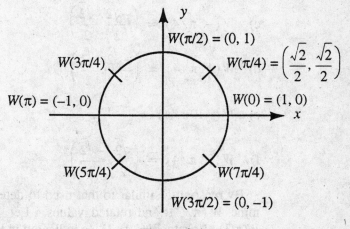

Figure 8.4

$W(-\pi/4)$ is the midpoint of the arc joining the points $W(0)$ and $W(3\pi/2)$. Since $W(\pi/4)$ and $W(-\pi/4)$ are symmetric with respect to the x-axis, we readily determine that

$$W(-\pi/4) = \left(\frac{\sqrt{2}}{2}, \frac{-\sqrt{2}}{2}\right).$$

Finally, $W(5\pi/4)$ is the midpoint of the arc joining the points $W(\pi)$ and $W(3\pi/2)$. Using symmetry and the coordinates for either $W(3\pi/4)$ or $W(-\pi/4)$, we can readily determine that

$$W(5\pi/4) = \left(\frac{-\sqrt{2}}{2}, \frac{-\sqrt{2}}{2}\right).$$

EXERCISE 8.3

Determine the function values for each of the following:

a) $W(-3\pi/4)$

b) $W(7\pi/4)$

c) $W(-5\pi/4)$

d) $W(11\pi/4)$

e) $W(-19\pi/4)$

f) $W(31\pi/4)$

SOLUTION 8.3

a) $W(-3\pi/4) = \left(\dfrac{-\sqrt{2}}{2}, \dfrac{-\sqrt{2}}{2}\right)$

b) $W(7\pi/4) = \left(\dfrac{\sqrt{2}}{2}, \dfrac{-\sqrt{2}}{2}\right)$

c) $W(-5\pi/4) = \left(\dfrac{-\sqrt{2}}{2}, \dfrac{\sqrt{2}}{2}\right)$

d) $W(11\pi/4) = \left(\dfrac{-\sqrt{2}}{2}, \dfrac{\sqrt{2}}{2}\right)$

e) $W(-19\pi/4) = \left(\dfrac{-\sqrt{2}}{2}, \dfrac{-\sqrt{2}}{2}\right)$

f) $W(31\pi/4) = \left(\dfrac{\sqrt{2}}{2}, \dfrac{-\sqrt{2}}{2}\right)$

By reasoning similar to that used to determine $W(\pi/4)$, we may determine $W(\pi/3)$ and related values. Let (x, y) be the coordinates of $W(\pi/3)$. Locate $W(2\pi/3)$ as indicated in Figure 8.5 and observe that

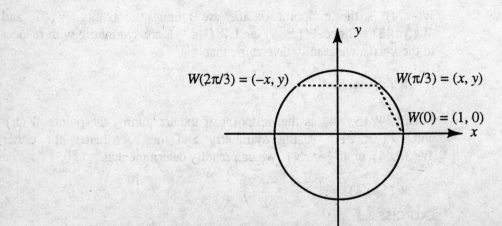

$W(2\pi/3) = (-x, y)$ $\qquad$ $W(\pi/3) = (x, y)$

$W(0) = (1, 0)$

Figure 8.5

$W(\pi/3)$ and $W(2\pi/3)$ are symmetric with respect to the y-axis. Hence, the coordinates of $W(2\pi/3)$ are $(-x, y)$. Since the length of the arc from $W(0)$ to $W(\pi/3)$ is equal to the length of the arc from $W(\pi/3)$ to $W(2\pi/3)$, the chords joining those points must also have the same lengths. Because these chords are symmetric with respect to the y-axis, the chord joining the points $W(2\pi/3)$ and $W(\pi/3)$ must have a measure of $x - (-x) = 2x$ units. The length of the chord from $W(0)$ to $W(\pi/3)$ may be found by using the undirected distance formula:

$$\sqrt{(x-1)^2 + (y-0)^2} = \sqrt{(x-1)^2 + y^2}$$

Since the two chords are equal in length, we have

$$2x = \sqrt{(x-1)^2 + y^2}$$

By squaring both sides of this equation, we obtain

$$4x^2 = (x-1)^2 + y^2$$

Simplifying, we get

$$4x^2 = x^2 - 2x + 1 + y^2$$
$$= (x^2 + y^2) - 2x + 1$$

Since $x^2 + y^2 = 1$, substitution for $x^2 + y^2$ in the last equation leads to

$$4x^2 = 1 - 2x + 1$$
$$= 2 - 2x.$$

Therefore,

$$4x^2 + 2x - 2 = 0$$

or

$$2(2x-1)(x+1) = 0$$

This last equation has two solutions,

$$x = 1/2 \quad \text{and} \quad x = -1$$

By examining Figure 8.5, we readily determine that $x = 1/2$ is the required solution. Since

$$y^2 = 1 - x^2,$$

we obtain

$$y^2 = 1 - \frac{1}{4}$$
$$= \frac{3}{4}$$

or

$$y = \frac{\sqrt{3}}{2} \qquad \text{(since } y \text{ must be positive).}$$

Therefore,

$$W(\pi/3) = \left(\frac{1}{2}, \frac{\sqrt{3}}{2}\right)$$

and

$$W(2\pi/3) = \left(\frac{-1}{2}, \frac{\sqrt{3}}{2}\right).$$

EXERCISE 8.4

Determine the function values for each of the following.

a) $W(-2\pi/3)$

b) $W(4\pi/3)$

c) $W(8\pi/3)$

d) $W(-5\pi/3)$

e) $W(10\pi/3)$

f) $W(-40\pi/3)$

SOLUTION 8.4

a) $W(-2\pi/3) = \left(\dfrac{-1}{2}, \dfrac{-\sqrt{3}}{2}\right)$

b) $W(4\pi/3) = \left(\dfrac{-1}{2}, \dfrac{-\sqrt{3}}{2}\right)$

c) $W(8\pi/3) = \left(\dfrac{-1}{2}, \dfrac{\sqrt{3}}{2}\right)$

d) $W(-5\pi/3) = \left(\dfrac{1}{2}, \dfrac{\sqrt{3}}{2}\right)$

e) $W(10\pi/3) = \left(\dfrac{-1}{2}, \dfrac{-\sqrt{3}}{2}\right)$

f) $W(-40\pi/3) = \left(\dfrac{-1}{2}, \dfrac{\sqrt{3}}{2}\right)$

The coordinates of $W(\pi/6)$ may be obtained in a manner similar to that used for $W(\pi/3)$. However, by examining Figure 8.6, we determine that the length of the arc from $W(0)$ to $W(\pi/6)$ is equal to the length of the arc from $W(\pi/3)$ to $W(\pi/2)$. Therefore, the distance between $W(\pi/6)$ and the x-axis is equal to the distance between $W(\pi/3)$ and the y-axis. It follows, then, that the x value of $W(\pi/6)$ is equal to the y value of $W(\pi/3)$, and that the y value of $W(\pi/6)$ is equal to the x value of $W(\pi/3)$. Since the coordinates of $W(\pi/3)$ were determined

to be $\left(\dfrac{1}{2}, \dfrac{\sqrt{3}}{2}\right)$, we readily determine that

$$W(\pi/6) = \left(\frac{\sqrt{3}}{2}, \frac{1}{2}\right).$$

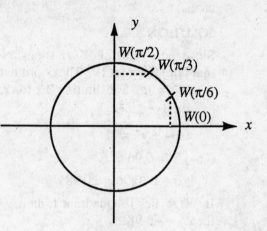

Figure 8.6

EXERCISE 8.5

Determine the function values for each of the following.
a) $W(5\pi/6)$
b) $W(-\pi/6)$
c) $W(11\pi/6)$
d) $W(-7\pi/6)$
e) $W(13\pi/6)$
f) $W(-31\pi/6)$

SOLUTION 8.5

a) $W(5\pi/6) = \left(\frac{-\sqrt{3}}{2}, \frac{1}{2}\right)$

b) $W(-\pi/6) = \left(\frac{\sqrt{3}}{2}, \frac{-1}{2}\right)$

c) $W(11\pi/6) = \left(\frac{\sqrt{3}}{2}, \frac{-1}{2}\right)$

d) $W(-7\pi/6) = \left(\frac{-\sqrt{3}}{2}, \frac{1}{2}\right)$

e) $W(13\pi/6) = \left(\frac{\sqrt{3}}{2}, \frac{1}{2}\right)$

f) $W(-31\pi/6) = \left(\frac{-\sqrt{3}}{2}, \frac{1}{2}\right)$

EXERCISE 8.6

If $W(t)$ has an x value of 0.2, determine its y value. (Hint: There are two cases.)

SOLUTION 8.6

Since the x value of $W(t)$ is positive, then $W(t)$ must lie in either Quadrant I or Quadrant IV. The coordinates of $W(t)$ must satisfy the equation $x^2 + y^2 = 1$. Substituting 0.2 for x, we have

$$(0.2)^2 + y^2 = 1$$
$$0.04 + y^2 = 1$$
$$y^2 = 0.96$$
$$y = \pm\sqrt{0.96} \approx \pm 0.98$$

If $W(t)$ lies in Quadrant I, then $y \approx 0.98$. If $W(t)$ lies in Quadrant IV, then $y \approx -0.98$.

EXERCISE 8.7

If $W(t)$ has a y value of -0.9, determine its x value. (Hint: There are two cases.)

SOLUTION 8.7

Since the y value of $W(t)$ is negative, then $W(t)$ must lie in either Quadrant III or Quadrant IV. The coordinates of $W(t)$ must satisfy the equation $x^2 + y^2 = 1$. Substituting -0.9 for y, we have

$$x^2 + y^2 = 1$$
$$x^2 + (-0.9)^2 = 1$$
$$x^2 + 0.81 = 1$$
$$x^2 = 0.19$$
$$x = \pm\sqrt{0.19} \approx \pm 0.44$$

If $W(t)$ lies in Quadrant III, then $x \approx -0.44$. If $W(t)$ lies in Quadrant IV, then $x \approx 0.44$.

8.2 THE SINE AND COSINE FUNCTIONS

From an analysis of the Wrapping function, we can readily obtain two additional functions. Observe that W maps each number on the number line onto a unique point on the circumference of the unit circle, C. But, each point on C has two coordinates—an x-coordinate and a y-coordinate. If t denotes an arbitrary element in the domain of W, then we see that as t

varies, x and y also vary. Further, each value of t is associated with a unique value of x and also a unique value of y. These associations lead to the following functions.

Definition:

Let W be the Wrapping function that maps each real number t on the number line onto a unique point (x, y) of the unit circle, C. We define two functions, **cosine** and **sine**, denoted by cos and sin, respectively, as follows:

$$x = \cos(t) = x\text{-coordinate of } W(t) \text{ for all } t \text{ in } (-\infty, +\infty)$$
$$y = \sin(t) = y\text{-coordinate of } W(t) \text{ for all } t \text{ in } (-\infty, +\infty)$$

We now make the following observations about the sine and cosine functions:

- The **domain** of both sin and cos is $(-\infty, +\infty)$.
- Since $-1 \leq x \leq 1$ and since x takes all of these values for various values of t, the **range** of cos is $[-1, 1]$.
- Similarly, since $-1 \leq y \leq 1$, and since y takes all of these values for various values of t, the **range** of sin is $[-1, 1]$.
- The sin and cos functions are **not one-to-one**. (This follows from the fact that the Wrapping function is not one-to-one.)
- The sin and cos functions do **not** have **inverses as functions**. (This also follows from the fact that the Wrapping function does not have an inverse as a function.)

Next, we attempt to generate function values for sin and cos. Observe that $W(t) = (x, y) = (\cos(t), \sin(t))$. Hence, it is fairly easy to determine some values for $\sin(t)$ and $\cos(t)$, although we shall have to use tables or calculators/computers for most of them. Since

$$W(0) = (1, 0)$$

then

$$\cos(0) = 1 \quad \text{and} \quad \sin(0) = 0.$$

In a similar manner, since

$$W(\pi/2) = (0, 1)$$

it follows that

$$\cos(\pi/2) = 0 \quad \text{and} \quad \sin(\pi/2) = 1.$$

From the fact that

$$W(\pi/4) = \left(\frac{\sqrt{2}}{2}, \frac{\sqrt{2}}{2}\right)$$

it follows that

$$\cos(\pi/4) = \frac{\sqrt{2}}{2} \quad \text{and} \quad \sin(\pi/4) = \frac{\sqrt{2}}{2}.$$

Continuing in this manner, we could easily compile a table of values for cos (*t*) and sin (*t*). (See Table 8.1.) Plotting the values for sin (*t*) and cos(*t*), as given in Table 8.1, and passing a smooth curve through them, we can graph the sin and cos functions from *t* = −2π to *t* = 2π. (See Figures 8.7 and 8.8.)

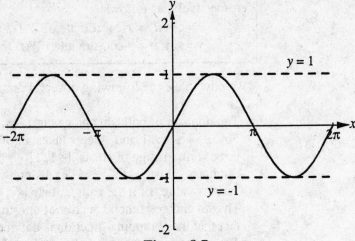

Figure 8.7
The Graph of $y = \sin(t)$ **for** $-2\pi < t < 2\pi$

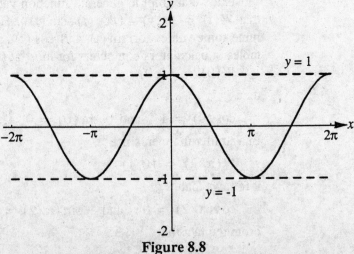

Figure 8.8
The Graph of $y = \cos(t)$ **for** $-2\pi < t < 2\pi$

t	$W(t)$	$\sin(t)$	$\cos(t)$	t	$W(t)$	$\sin(t)$	$\cos(t)$
0	$(1, 0)$	0	1				
$\dfrac{\pi}{6}$	$\left(\dfrac{\sqrt{3}}{2}, \dfrac{1}{2}\right)$	$\dfrac{1}{2}$	$\dfrac{\sqrt{3}}{2}$	$-\dfrac{\pi}{6}$	$\left(\dfrac{\sqrt{3}}{2}, -\dfrac{1}{2}\right)$	$-\dfrac{1}{2}$	$\dfrac{\sqrt{3}}{2}$
$\dfrac{\pi}{4}$	$\left(\dfrac{\sqrt{2}}{2}, \dfrac{\sqrt{2}}{2}\right)$	$\dfrac{\sqrt{2}}{2}$	$\dfrac{\sqrt{2}}{2}$	$-\dfrac{\pi}{4}$	$\left(\dfrac{\sqrt{2}}{2}, -\dfrac{\sqrt{2}}{2}\right)$	$-\dfrac{\sqrt{2}}{2}$	$\dfrac{\sqrt{2}}{2}$
$\dfrac{\pi}{3}$	$\left(\dfrac{1}{2}, \dfrac{\sqrt{3}}{2}\right)$	$\dfrac{\sqrt{3}}{2}$	$\dfrac{1}{2}$	$-\dfrac{\pi}{3}$	$\left(\dfrac{1}{2}, -\dfrac{\sqrt{3}}{2}\right)$	$-\dfrac{\sqrt{3}}{2}$	$\dfrac{1}{2}$
$\dfrac{\pi}{2}$	$(0, 1)$	1	0	$-\dfrac{\pi}{2}$	$(0, -1)$	-1	0
$\dfrac{2\pi}{3}$	$\left(-\dfrac{1}{2}, \dfrac{\sqrt{3}}{2}\right)$	$\dfrac{\sqrt{3}}{2}$	$-\dfrac{1}{2}$	$-\dfrac{2\pi}{3}$	$\left(-\dfrac{1}{2}, -\dfrac{\sqrt{3}}{2}\right)$	$-\dfrac{\sqrt{3}}{2}$	$-\dfrac{1}{2}$
$\dfrac{3\pi}{4}$	$\left(-\dfrac{\sqrt{2}}{2}, \dfrac{\sqrt{2}}{2}\right)$	$\dfrac{\sqrt{2}}{2}$	$-\dfrac{\sqrt{2}}{2}$	$-\dfrac{3\pi}{4}$	$\left(-\dfrac{\sqrt{2}}{2}, -\dfrac{\sqrt{2}}{2}\right)$	$-\dfrac{\sqrt{2}}{2}$	$-\dfrac{\sqrt{2}}{2}$
$\dfrac{5\pi}{6}$	$\left(-\dfrac{\sqrt{3}}{2}, \dfrac{1}{2}\right)$	$\dfrac{1}{2}$	$-\dfrac{\sqrt{3}}{2}$	$-\dfrac{5\pi}{6}$	$\left(-\dfrac{\sqrt{3}}{2}, -\dfrac{1}{2}\right)$	$-\dfrac{1}{2}$	$-\dfrac{\sqrt{3}}{2}$
π	$(-1, 0)$	0	-1	$-\pi$	$(-1, 0)$	0	-1
$\dfrac{7\pi}{6}$	$\left(-\dfrac{\sqrt{3}}{2}, -\dfrac{1}{2}\right)$	$-\dfrac{1}{2}$	$-\dfrac{\sqrt{3}}{2}$	$-\dfrac{7\pi}{6}$	$\left(-\dfrac{\sqrt{3}}{2}, \dfrac{1}{2}\right)$	$\dfrac{1}{2}$	$-\dfrac{\sqrt{3}}{2}$
$\dfrac{5\pi}{4}$	$\left(-\dfrac{\sqrt{2}}{2}, -\dfrac{\sqrt{2}}{2}\right)$	$-\dfrac{\sqrt{2}}{2}$	$-\dfrac{\sqrt{2}}{2}$	$-\dfrac{5\pi}{4}$	$\left(-\dfrac{\sqrt{2}}{2}, \dfrac{\sqrt{2}}{2}\right)$	$\dfrac{\sqrt{2}}{2}$	$-\dfrac{\sqrt{2}}{2}$
$\dfrac{4\pi}{3}$	$\left(-\dfrac{1}{2}, -\dfrac{\sqrt{3}}{2}\right)$	$-\dfrac{\sqrt{3}}{2}$	$-\dfrac{1}{2}$	$-\dfrac{4\pi}{3}$	$\left(-\dfrac{1}{2}, \dfrac{\sqrt{3}}{2}\right)$	$\dfrac{\sqrt{3}}{2}$	$-\dfrac{1}{2}$
$\dfrac{3\pi}{2}$	$(0, -1)$	-1	0	$-\dfrac{3\pi}{2}$	$(0, 1)$	1	0
$\dfrac{5\pi}{3}$	$\left(\dfrac{1}{2}, -\dfrac{\sqrt{3}}{2}\right)$	$-\dfrac{\sqrt{3}}{2}$	$\dfrac{1}{2}$	$-\dfrac{5\pi}{3}$	$\left(\dfrac{1}{2}, \dfrac{\sqrt{3}}{2}\right)$	$\dfrac{\sqrt{3}}{2}$	$\dfrac{1}{2}$
$\dfrac{7\pi}{4}$	$\left(\dfrac{\sqrt{2}}{2}, -\dfrac{\sqrt{2}}{2}\right)$	$-\dfrac{\sqrt{2}}{2}$	$\dfrac{\sqrt{2}}{2}$	$-\dfrac{7\pi}{4}$	$\left(\dfrac{\sqrt{2}}{2}, \dfrac{\sqrt{2}}{2}\right)$	$\dfrac{\sqrt{2}}{2}$	$\dfrac{\sqrt{2}}{2}$
$\dfrac{11\pi}{6}$	$\left(\dfrac{\sqrt{3}}{2}, -\dfrac{1}{2}\right)$	$-\dfrac{1}{2}$	$\dfrac{\sqrt{3}}{2}$	$-\dfrac{11\pi}{6}$	$\left(\dfrac{\sqrt{3}}{2}, \dfrac{1}{2}\right)$	$\dfrac{1}{2}$	$\dfrac{\sqrt{3}}{2}$
2π	$(1, 0)$	0	1	-2π	$(1, 0)$	0	1

Table 8.1
Partial Table of Values for sin(t) and cos(t)

EXERCISE 8.8

Determine $\sin(t)$ and $\cos(t)$ for the indicated values of t.

a) $t = 6\pi$

b) $t = 5\pi/2$

c) $t = -7\pi/2$

d) $t = -9\pi$

e) $t = 51\pi$

f) $t = -21\pi/2$

g) $t = 19\pi/2$

SOLUTION 8.8

a) $\sin(6\pi) = y$-coordinate of $W(6\pi) = 0$.

   $\cos(6\pi) = x$-coordinate of $W(6\pi) = 1$.

b) $\sin(5\pi/2) = y$-coordinate of $W(5\pi/2) = 1$.

   $\cos(5\pi/2) = x$-coordinate of $W(5\pi/2) = 0$.

c) $\sin(-7\pi/2) = y$-coordinate of $W(-7\pi/2) = 1$.

   $\cos(-7\pi/2) = x$-coordinate of $W(-7\pi/2) = 0$.

d) $\sin(-9\pi) = y$-coordinate of $W(-9\pi) = 0$.

   $\cos(-9\pi) = x$-coordinate of $W(-9\pi) = -1$.

e) $\sin(51\pi) = y$-coordinate of $W(51\pi) = 0$.

   $\cos(51\pi) = x$-coordinate of $W(51\pi) = -1$.

f) $\sin(-21\pi/2) = y$-coordinate of $W(-21\pi/2) = -1$.

   $\cos(-21\pi/2) = x$-coordinate of $W(-21\pi/2) = 0$.

g) $\sin(19\pi/2) = y$-coordinate of $W(19\pi/2) = -1$.

   $\cos(19\pi/2) = x$-coordinate of $W(19\pi/2) = 0$.

EXERCISE 8.9

Determine $\sin(t)$ and $\cos(t)$ for the indicated values of t.

a) $t = -3\pi/4$

b) $t = 7\pi/4$

c) $t = 23\pi/4$

d) $t = -19\pi/4$

e) $t = -5\pi/4$

f) $t = 31\pi/4$

SOLUTION 8.9

a) $\sin(-3\pi/4) = y$-coordinate of $W(-3\pi/4) = \dfrac{-\sqrt{2}}{2}$.

$\cos(-3\pi/4) = x$-coordinate of $W(-3\pi/4) = \dfrac{-\sqrt{2}}{2}$.

b) $\sin(7\pi/4) = y$-coordinate of $W(7\pi/4) = \dfrac{-\sqrt{2}}{2}$.

$\cos(7\pi/4) = x$-coordinate of $W(7\pi/4) = \dfrac{\sqrt{2}}{2}$.

c) $\sin(23\pi/4) = y$-coordinate of $W(23\pi/4) = \dfrac{-\sqrt{2}}{2}$.

$\cos(23\pi/4) = x$-coordinate of $W(23\pi/4) = \dfrac{\sqrt{2}}{2}$.

d) $\sin(-19\pi/4) = y$-coordinate of $W(-19\pi/4) = \dfrac{-\sqrt{2}}{2}$.

$\cos(-19\pi/4) = x$-coordinate of $W(-19\pi/4) = \dfrac{-\sqrt{2}}{2}$.

e) $\sin(-5\pi/4) = y$-coordinate of $W(-5\pi/4) = \dfrac{\sqrt{2}}{2}$.

$\cos(-5\pi/4) = x$-coordinate of $W(-5\pi/4) = \dfrac{-\sqrt{2}}{2}$.

f) $\sin(31\pi/4) = y$-coordinate of $W(31\pi/4) = \dfrac{-\sqrt{2}}{2}$.

$\cos(31\pi/4) = x$-coordinate of $W(31\pi/4) = \dfrac{\sqrt{2}}{2}$.

EXERCISE 8.10

Determine $\sin(t)$ and $\cos(t)$ for the indicated values of t.

a) $t = -2\pi/3$

b) $t = 4\pi/3$

c) $t = 20\pi/3$

d) $t = -40\pi/3$

e) $t = -8\pi/3$

f) $t = 14\pi/3$

SOLUTION 8.10

a) $\sin(-2\pi/3)$ = y-coordinate of $W(-2\pi/3)$ = $\dfrac{-\sqrt{3}}{2}$.

 $\cos(-2\pi/3)$ = x-coordinate of $W(-2\pi/3)$ = $\dfrac{-1}{2}$.

b) $\sin(4\pi/3)$ = y-coordinate of $W(4\pi/3)$ = $\dfrac{-\sqrt{3}}{2}$.

 $\cos(4\pi/3)$ = x-coordinate of $W(4\pi/3)$ = $\dfrac{-1}{2}$.

c) $\sin(20\pi/3)$ = y-coordinate of $W(20\pi/3)$ = $\dfrac{\sqrt{3}}{2}$.

 $\cos(20\pi/3)$ = x-coordinate of $W(20\pi/3)$ = $\dfrac{-1}{2}$.

d) $\sin(-40\pi/3)$ = y-coordinate of $W(-40\pi/3)$ = $\dfrac{\sqrt{3}}{2}$.

 $\cos(-40\pi/3)$ = x-coordinate of $W(-40\pi/3)$ = $\dfrac{-1}{2}$.

e) $\sin(-8\pi/3)$ = y-coordinate of $W(-8\pi/3)$ = $\dfrac{-\sqrt{3}}{2}$.

 $\cos(-8\pi/3)$ = x-coordinate of $W(-8\pi/3)$ = $\dfrac{-1}{2}$.

f) $\sin(14\pi/3)$ = y-coordinate of $W(14\pi/3)$ = $\dfrac{\sqrt{3}}{2}$.

 $\cos(14\pi/3)$ = x-coordinate of $W(14\pi/3)$ = $\dfrac{-1}{2}$.

EXERCISE 8.11

Determine $\sin(t)$ and $\cos(t)$ for the indicated values of t.

a) $t = 7\pi/6$

b) $t = -5\pi/6$

c) $t = 11\pi/6$

d) $t = -13\pi/6$

e) $t = 31\pi/6$

f) $t = -50\pi/6$

SOLUTION 8.11

a) Since $W(7\pi/6) = \left(\dfrac{-\sqrt{3}}{2}, \dfrac{-1}{2}\right)$, $\sin(7\pi/6) = \dfrac{-1}{2}$ and $\cos(7\pi/6) = \dfrac{-\sqrt{3}}{2}$.

b) Since $W(-5\pi/6) = \left(\dfrac{-\sqrt{3}}{2}, \dfrac{-1}{2}\right)$, $\sin(-5\pi/6) = \dfrac{-1}{2}$ and $\cos(-5\pi/6) = \dfrac{-\sqrt{3}}{2}$.

c) Since $W(11\pi/6) = \left(\dfrac{-\sqrt{3}}{2}, \dfrac{-1}{2}\right)$, $\sin(11\pi/6) = \dfrac{-1}{2}$ and $\cos(11\pi/6) = \dfrac{\sqrt{3}}{2}$.

d) Since $W(-13\pi/6) = \left(\dfrac{\sqrt{3}}{2}, \dfrac{-1}{2}\right)$, $\sin(-13\pi/6) = \dfrac{-1}{2}$ and $\cos(-13\pi/6) = \dfrac{\sqrt{3}}{2}$.

e) Since $W(31\pi/6) = \left(\dfrac{-\sqrt{3}}{2}, \dfrac{-1}{2}\right)$, $\sin(31\pi/6) = \dfrac{-1}{2}$ and $\cos(31\pi/6) = \dfrac{-\sqrt{3}}{2}$.

f) Since $W(-50\pi/6) = \left(\dfrac{1}{2}, \dfrac{-\sqrt{3}}{2}\right)$, $\sin(-50\pi/6) = \dfrac{-\sqrt{3}}{2}$ and $\cos(-50\pi/6) = \dfrac{1}{2}$.

EXERCISE 8.12

Determine whether the given function is increasing or decreasing on the indicated interval.

a) sin on $[\pi/2, \pi]$

b) cos on $[-\pi/2, 0]$

c) sin on $[3\pi/2, 2\pi]$

d) cos on $[-2\pi, -3\pi/2]$

e) sin on $[9\pi/2, 5\pi]$

f) cos on $[-7\pi/2, -3\pi]$

SOLUTION 8.12

a) Decreasing

b) Increasing

c)   Increasing

d)   Decreasing

e)   Decreasing

f)   Decreasing

Periodic Functions   From Figure 8.7, we note that the portion of the graph of sin for $0 \le t \le 2\pi$ is identical to the portion of the graph for $-2\pi \le t \le 0$. The same observation is true for the graph of cos. In fact, if we continued the graphs for $2\pi \le t \le 4\pi$, they would again duplicate the portion of the graphs for $0 \le t \le 2\pi$. Such functions are said to be **periodic**.

Definition:

The function F is said to be **periodic, with period k**, if

$F(x+k) = F(x)$ for *all* x in the domain of F.

Both the sin and cos functions are periodic, with a period of 2π. We have

$\sin(t+2\pi) = \sin(t)$

$\sin[(t+2\pi)+2\pi] = \sin(t+2\pi) = \sin(t)$

In particular,

$\sin(t+2\pi k) = \sin(t)$ for $k = 0, \pm1, \pm2, \pm3, \ldots$

Also,

$\cos(t+2\pi k) = \cos(t)$ for $k = 0, \pm1, \pm2, \pm3, \ldots$

Since both sin and cos are periodic functions with period 2π, their graphs repeat every 2π units.

8.3 INVERSE SINE AND INVERSE COSINE FUNCTIONS

The domain of the Sine function is the set $(-\infty, +\infty)$ and the range is the set $[-1, 1]$. The Sine function is periodic with a period 2π. Hence, the Sine function is not one-to-one. For each value in the range of the Sine function, there exist infinitely many values in its domain mapped onto it. However, if we place a suitable restriction on the domain of the function, we can construct a function that will be one-to-one and, consequently, will have an inverse as a function. There are different ways of restricting the domain to accomplish this, but we shall use the one that has been most commonly accepted by mathematicians.

The Restricted Sine Function and Its Inverse

From the graph of the Sine function given in Figure 8.7, we determine that if the domain of the function is restricted to the interval $[-\pi/2, \pi/2]$, the **restricted Sine function** will have an inverse as a function. This inverse function is denoted by $\sin^{-1}$

Definition:

Consider the *restricted* Sine function with domain $[-\pi/2, \pi/2]$. The **Inverse Sine** function, denoted by $\sin^{-1}$, is defined by $y = \sin^{-1}(x)$, if and only if $x = \sin(y)$. The symbol $\sin^{-1}(x)$ is read "the inverse sine of x."

We now make some observations about the Inverse Sine function.

- The **domain** of $\sin^{-1}$ is the interval $[-1, 1]$, which is equal to the range of the restricted Sine function.
- The **range** of $\sin^{-1}$ is the interval $[-\pi/2, \pi/2]$, which is equal to the domain of the restricted Sine function.
- The graph of $\sin^{-1}$ is obtained by reflecting the graph of sin (restricted to $[-\pi/2, \pi/2]$) about the line with equation $y = x$. (The graphs of the restricted sine function and $\sin^{-1}$ are given in Figure 8.9.)
- The $\sin^{-1}$ function is **one-to-one**.
- The $\sin^{-1}$ function has an **inverse as a function**. The inverse function is the Sine function restricted to $[-\pi/2, \pi/2]$.

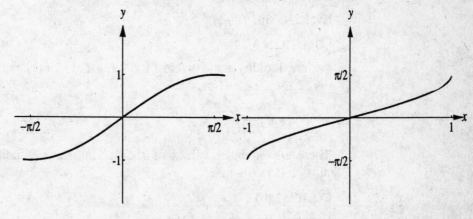

Figure 8.9

a) **The Graph of $y = \sin(x)$ restricted to $[-\pi/2, \pi/2]$**   b) **The Graph of $y = \sin^{-1}(x)$**

Some authors refer to the inverse Sine function as the Arcsine function and use the symbol Arcsin instead of $\sin^{-1}$. We shall not use that notation in this text but merely wish to call it to the student's attention.

EXERCISE 8.13

Evaluate $\sin^{-1}(1/2)$.

SOLUTION 8.13

We are looking for a value of t such that $\sin(t) = 1/2$ and such that $-\pi/2 \le t \le \pi/2$. But,

$$\sin(\pi/6) = 1/2 \qquad \text{and} \qquad -\pi/2 \le \pi/6 \le \pi/2$$

There are no other values of t satisfying these conditions. Therefore, $\sin^{-1}(1/2) = \pi/6$.

EXERCISE 8.14

Evaluate $\sin^{-1}(-\sqrt{3}/2)$.

SOLUTION 8.14

We are looking for a value of t such that $\sin(t) = -\sqrt{3}/2$ and such that $-\pi/2 \le t \le \pi/2$. But,

$$\sin(-\pi/3) = -\sqrt{3}/2 \qquad \text{and} \qquad -\pi/2 \le -\pi/3 \le \pi/2$$

There are no other values of t satisfying these conditions. Therefore, $\sin^{-1}(-\sqrt{3}/2) = -\pi/3$.

EXERCISE 8.15

Evaluate $\sin^{-1}(-1)$.

SOLUTION 8.15

We are looking for a value of t such that $\sin(t) = -1$ and such that $-\pi/2 \le t \le \pi/2$. But,

$$\sin(-\pi/2) = -1 \qquad \text{and} \qquad -\pi/2 \le -\pi/2 \le \pi/2.$$

There are no other values of t satisfying these conditions. Therefore, $\sin^{-1}(-1) = -\pi/2$.

EXERCISE 8.16

Evaluate $\sin(\sin^{-1}(1/2))$.

SOLUTION 8.16

$$\sin(\sin^{-1}(1/2)) = \sin(\pi/6) \qquad \text{(from Exercise 8.13)}$$
$$= 1/2$$

Therefore, $\sin(\sin^{-1}(1/2)) = 1/2$.

Note that in the above exercise we have the composition of the functions sin and sin^{-1} which is equal to the Identity function.

The Restricted Cosine Function and Its Inverse

We can also restrict the Cosine function to a proper subset of its domain and construct a new function that will have an inverse.

> **Definition:**
> Consider the restricted cosine function with domain $[0, \pi]$. The **Inverse Cosine** function, denoted by $\cos^{-1}$, is defined by
> $y = \cos^{-1}(x)$, if and only if $x = \cos(y)$. The symbol $\cos^{-1}(x)$ is read "the inverse cosine of x."

We now make some observations about the Inverse Cosine function.

- The **domain** of $\cos^{-1}$ is $[-1, 1]$, which is equal to the range of the restricted Cosine function.
- The **range** of $\cos^{-1}$ is $[0, \pi]$, which is equal to the domain of the restricted cosine function.
- The graph of $\cos^{-1}$ is obtained by reflecting the graph of cos (restricted to $[0, \pi]$) about the line with equation $y = x$. (The graphs of the restricted cosine function and $\cos^{-1}$ are given in Figure 8.10.)
- The $\cos^{-1}$ function is **one-to-one**.
- The $\cos^{-1}$ function has an **inverse as a function**. The inverse function is the cosine function restricted to $[0, \pi]$.

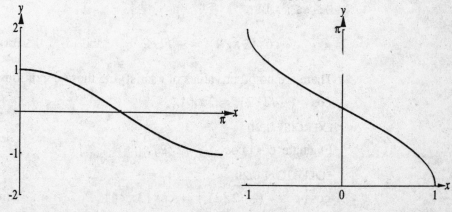

Figure 8.10

a) The Graph of $y = \cos(x)$ restricted to $[0, \pi]$

b) The Graph of $y = \cos^{-1}(x)$

EXERCISE 8.17

Evaluate $\cos^{-1}(-1/2)$.

SOLUTION 8.17

We are looking for a value of t such that $\cos(t) = -1/2$ and such that $0 \le t \le \pi$. But,

$$\cos(2\pi/3) = -1/2 \qquad \text{and} \qquad 0 \le 2\pi/3 \le \pi.$$

There are no other values of t satisfying these conditions. Therefore, $\cos^{-1}(-1/2) = 2\pi/3$.

EXERCISE 8.18

Evaluate $\cos^{-1}(0)$.

SOLUTION 8.18

We are looking for a value of t such that $\cos(t) = 0$ and such that $0 \le t \le \pi$. But,

$$\cos(\pi/2) = 0 \qquad \text{and} \qquad 0 \le \pi/2 \le \pi.$$

There are no other values of t satisfying these conditions. Therefore, $\cos^{-1}(0) = \pi/2$.

EXERCISE 8.19

Evaluate $\cos^{-1}(-\sqrt{2}/2)$.

SOLUTION 8.19

We are looking for a value of t such that $\cos(t) = -\sqrt{2}/2$ and such that $0 \le t \le \pi$. But,

$$\cos(3\pi/4) = -\sqrt{2}/2 \qquad \text{and} \qquad 0 \le 3\pi/4 \le \pi.$$

There are no other values of t satisfying these conditions. Therefore, $\cos^{-1}(-\sqrt{2}/2) = 3\pi/4$.

EXERCISE 8.20

Evaluate $\cos\left(\cos^{-1}(-\sqrt{2}/2)\right)$.

SOLUTION 8.20

$$\cos\left(\cos^{-1}(-\sqrt{2}/2)\right) = \cos(3\pi/4) \qquad \text{(from Exercise 8.19)}$$
$$= -\sqrt{2}/2$$

EXERCISE 8.21

Evaluate $\sin\left(\cos^{-1}(-\sqrt{3}/2)\right)$.

SOLUTION 8.21

Let $t = \cos^{-1}(-\sqrt{3}/2)$. Then,

$$\cos(t) = -\sqrt{3}/2 \quad \text{and} \quad 0 \le t \le \pi.$$

We determine that $t = 5\pi/6$ will work and that there are no other values of t that satisfy the above conditions. Hence,

$$\sin\left(\cos^{-1}(-\sqrt{3}/2)\right) = \sin(5\pi/6)$$
$$= 1/2$$

Therefore, $\sin\left(\cos^{-1}(-\sqrt{3}/2)\right) = 1/2$.

EXERCISE 8.22

Evaluate $\sin(2\cos^{-1}(1/2))$.

SOLUTION 8.22

$\cos^{-1}(1/2)$ is equivalent to $\cos(t) = 1/2$ for $0 \le t \le \pi$.

We determine that $t = \pi/3$. Hence,

$$\sin(2\cos^{-1}(1/2)) = \sin(2 \cdot \pi/3)$$
$$= \sin(2\pi/3)$$
$$= \sqrt{3}/2$$

Therefore, $\sin(2\cos^{-1}(1/2)) = \sqrt{3}/2$.

EXERCISE 8.23

Evaluate $5\cos^{-1}(\sin(0))$.

SOLUTION 8.23

$$5\cos^{-1}(\sin(0)) = 5\cos^{-1}(0) \qquad (\text{Since } \sin(0) = 0)$$
$$= 5(\pi/2) \qquad (\text{Since } \cos^{-1}(0) = \pi/2)$$
$$= 5\pi/2$$

Therefore, $5\cos^{-1}(\sin(0)) = 5\pi/2$.

EXERCISE 8.24

Evaluate $\cos^{-1}(1) - 3\sin^{-1}(-\sqrt{3}/2)$.

SOLUTION 8.24

Let $t = \cos^{-1}(1)$. Then, $\cos(t) = 1$ for $0 \le t \le \pi$. We determine that $t = 0$.

Let $\sin^{-1}(-\sqrt{3}/2) = p$. Then, $\sin(p) = -\sqrt{3}/2$ for $-\pi/2 \le p \le \pi/2$. We determine that $p = -\pi/3$. Hence,

$$\cos^{-1}(1) - 3\sin^{-1}(-\sqrt{3}/2) = 0 - 3(-\pi/3)$$
$$= \pi.$$

Therefore, $\cos^{-1}(1) - 3\sin^{-1}(-\sqrt{3}/2) = \pi$.

EXERCISE 8.25

Evaluate $4\cos^{-1}(-1) - 2\sin^{-1}(-\sqrt{2}/2)$.

SOLUTION 8.25

Let $t = \cos^{-1}(-1)$. Then, $\cos(t) = -1$ for $0 \le t \le \pi$. We determine that $t = \pi$.

Let $p = \sin^{-1}(-\sqrt{2}/2)$. Then, $\sin(p) = -\sqrt{2}/2$ for $-\pi/2 \le p \le \pi/2$. We determine that $p = -\pi/4$. Hence,

$$4 \cdot \cos^{-1}(-1) - \left(2 \cdot \sin^{-1}(-\sqrt{2}/2)\right) = 4(\pi) - 2(-\pi/4)$$
$$= 4\pi + \pi/2$$
$$= 9\pi/2.$$

Therefore, $4 \cdot \cos^{-1}(-1) - 2\sin^{-1}(-\sqrt{2}/2) = 9\pi/2$.

8.4 TANGENT AND COTANGENT FUNCTIONS AND THEIR INVERSES

Additional circular functions, together with the inverses of the restricted functions, will now be introduced, using the algebra of functions. Consider the functions sin and cos and form the quotients $\frac{\sin}{\cos}$ and $\frac{\cos}{\sin}$. With suitable qualifiers, new circular functions are defined.

Tangent and Cotangent Functions

> **Definition:**
> Let sin and cos be the sine and cosine functions, each with domain $(-\infty, +\infty)$. The **Tangent** function, denoted by tan, is defined by
> $$y = \tan(x) = \frac{\sin(x)}{\cos(x)}$$
> such that x is a real number and $\cos(x) \ne 0$.

From an examination of the Cosine function, we know that $\cos(x) = 0$ if $x = \pi/2$ or $x = (2n+1)\pi/2$, where n is any *integer*. Hence, the domain of the Tangent function is the set of all real numbers that are not odd integer multiples of $\pi/2$. The range of the Tangent function is $(-\infty, +\infty)$.

Definition:

Let sin and cos be the sine and cosine functions, each with domain $(-\infty, +\infty)$. The **Cotangent** function, denoted by cot, is defined by

$$y = \cot(x) = \frac{\cos(x)}{\sin(x)}$$

such that x is a real number and $\sin(x) \neq 0$.

From an examination of the Sine function, we know that $\sin(x) = 0$ if $x = 0$ or $x = n\pi$, where n is any *integer*. Hence, the domain of the cotangent function is the set of all real numbers that are not integer multiples of π. The range of the cotangent function is $(-\infty, +\infty)$.

We now make the following observations about the Tangent and cotangent functions.

- The **domain** of the tan function is the set of all **real numbers** that are **not odd integer multiples of $\pi/2$.**
- The **domain** of the cot function is the set of all **real numbers** that are **not integer multiples of π.**
- The **range** of both the tan and cot functions is $(-\infty, +\infty)$.
- **Neither** the tan nor the cot function is **one-to-one.**
- **Neither** the tan nor the cot function has an **inverse as a function.**
- Both the tan and cot functions are **periodic** with period π.

EXERCISE 8.26

Evaluate.

a) $\tan(\pi/6)$

b) $\cot(\pi/6)$

SOLUTION 8.26

Since $\sin(\pi/6) = 1/2$ and $\cos(\pi/6) = \sqrt{3}/2$, we have:

a) $\tan(\pi/6) = \dfrac{\sin(\pi/6)}{\cos(\pi/6)} = \dfrac{1/2}{\sqrt{3}/2} = \dfrac{1}{\sqrt{3}} = \dfrac{\sqrt{3}}{3}$

b) $\cot(\pi/6) = \dfrac{\cos(\pi/6)}{\sin(\pi/6)} = \dfrac{\sqrt{3}/2}{1/2} = \dfrac{\sqrt{3}}{1} = \sqrt{3}$

EXERCISE 8.27

Evaluate

a) $\tan(-7\pi/2)$

b) $\cot(-7\pi/2)$

SOLUTION 8.27

a) $\tan(-7\pi/2) = \dfrac{\sin(-7\pi/2)}{\cos(-7\pi/2)} = \dfrac{1}{0}$, which is not defined.

b) $\cot(-7\pi/2) = \dfrac{\cos(-7\pi/2)}{\sin(-7\pi/2)} = \dfrac{0}{1} = 0$

EXERCISE 8.28

Evaluate

a) $\tan(9\pi)$

b) $\cot(9\pi)$

SOLUTION 8.28

a) $\tan(9\pi) = \dfrac{\sin(9\pi)}{\cos(9\pi)} = \dfrac{0}{-1} = 0$

b) $\cot(9\pi) = \dfrac{\cos(9\pi)}{\sin(9\pi)} = \dfrac{-1}{0}$, which is not defined.

EXERCISE 8.29

Evaluate

a) $\tan(23\pi/4)$

b) $\cot(23\pi/4)$

SOLUTION 8.29

a) $\tan(23\pi/4) = \dfrac{\sin(23\pi/4)}{\cos(23\pi/4)} = \dfrac{-\sqrt{2}/2}{\sqrt{2}/2} = -1$

b) $\cot(23\pi/4) = \dfrac{\cos(23\pi/4)}{\sin(23\pi/4)} = \dfrac{\sqrt{2}/2}{-\sqrt{2}/2} = -1$

EXERCISE 8.30

Evaluate

a) $\tan(8\pi/3)$

b) $\cot(8\pi/3)$

SOLUTION 8.30

a) $\tan(8\pi/3) = \dfrac{\sin(8\pi/3)}{\cos(8\pi/3)} = \dfrac{\sqrt{3}/2}{-1/2} = -\sqrt{3}$

b) $\cot(8\pi/3) = \dfrac{\cos(8\pi/3)}{\sin(8\pi/3)} = \dfrac{-1/2}{\sqrt{3}/2} = \dfrac{-1}{\sqrt{3}}$

EXERCISE 8.31

Given that $\tan(t) > 0$, determine the quadrant(s) for $W(t)$.

SOLUTION 8.31

Since $\tan(t) = \dfrac{\sin(t)}{\cos(t)}$ provided that $\cos(t) \neq 0$, and since $\tan(t) > 0$, there are two possibilities. Either $\sin(t) > 0$ **and** $\cos(t) > 0$, *or* $\sin(t) < 0$ **and** $\cos(t) < 0$. Now, if $\sin(t) > 0$ **and** $\cos(t) > 0$, then $W(t)$ must be in Quadrant I. However, if $\sin(t) < 0$ **and** $\cos(t) < 0$, then $W(t)$ must be in Quadrant III. Hence, if $\tan(t) > 0$, then $W(t)$ is either in Quadrant I or in Quadrant III.

We can continue to evaluate values for $\tan(t)$ and $\cot(t)$ using the procedure indicated in the previous exercises. However, we have a partial table of such values, given in Table 8.2.

Plotting the values of $\tan(t)$ and $\cot(t)$ given in Table 8.2 and passing a smooth curve through them, we obtain the graphs of the tan and cot functions from $t = -2\pi$ to $t = 2\pi$. (See Figures 8.11 and 8.12.) From the graphs given in Figures 8.11 and 8.12, we determine that both the tan and cot functions are periodic, with period π. Further, we determine that neither of these functions possesses an inverse as a function. However, with suitable restrictions of the domain, each function will have an inverse function.

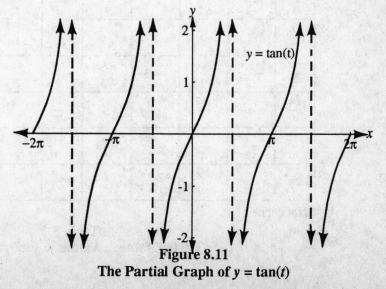

Figure 8.11
The Partial Graph of $y = \tan(t)$

t	$\sin(t)$	$\cos(t)$	$\tan(t)$	$\cot(t)$	t	$\sin(t)$	$\cos(t)$	$\tan(t)$	$\cot(t)$
0	0	1	0	$*$					
$\dfrac{\pi}{6}$	$\dfrac{1}{2}$	$\dfrac{\sqrt{3}}{2}$	$\dfrac{\sqrt{3}}{3}$	$\sqrt{3}$	$-\dfrac{\pi}{6}$	$-\dfrac{1}{2}$	$\dfrac{\sqrt{3}}{2}$	$-\dfrac{\sqrt{3}}{3}$	$-\sqrt{3}$
$\dfrac{\pi}{4}$	$\dfrac{\sqrt{2}}{2}$	$\dfrac{\sqrt{2}}{2}$	1	1	$-\dfrac{\pi}{4}$	$-\dfrac{\sqrt{2}}{2}$	$\dfrac{\sqrt{2}}{2}$	-1	-1
$\dfrac{\pi}{3}$	$\dfrac{\sqrt{3}}{2}$	$\dfrac{1}{2}$	$\sqrt{3}$	$\dfrac{\sqrt{3}}{3}$	$-\dfrac{\pi}{3}$	$-\dfrac{\sqrt{3}}{2}$	$\dfrac{1}{2}$	$-\sqrt{3}$	$-\dfrac{\sqrt{3}}{3}$
$\dfrac{\pi}{2}$	1	0	$*$	0	$-\dfrac{\pi}{2}$	-1	0	$*$	0
$\dfrac{2\pi}{3}$	$\dfrac{\sqrt{3}}{2}$	$-\dfrac{1}{2}$	$-\sqrt{3}$	$-\dfrac{\sqrt{3}}{3}$	$-\dfrac{2\pi}{3}$	$-\dfrac{\sqrt{3}}{2}$	$-\dfrac{1}{2}$	$\sqrt{3}$	$\dfrac{\sqrt{3}}{3}$
$\dfrac{3\pi}{4}$	$\dfrac{\sqrt{2}}{2}$	$-\dfrac{\sqrt{2}}{2}$	-1	-1	$-\dfrac{3\pi}{4}$	$-\dfrac{\sqrt{2}}{2}$	$-\dfrac{\sqrt{2}}{2}$	1	1
$\dfrac{5\pi}{6}$	$\dfrac{1}{2}$	$-\dfrac{\sqrt{3}}{2}$	$-\dfrac{\sqrt{3}}{3}$	$-\sqrt{3}$	$-\dfrac{5\pi}{6}$	$-\dfrac{1}{2}$	$-\dfrac{\sqrt{3}}{2}$	$\dfrac{\sqrt{3}}{3}$	$\sqrt{3}$
π	0	-1	0	$*$	$-\pi$	0	-1	0	$*$
$\dfrac{7\pi}{6}$	$-\dfrac{1}{2}$	$-\dfrac{\sqrt{3}}{2}$	$\dfrac{\sqrt{3}}{3}$	$\sqrt{3}$	$-\dfrac{7\pi}{6}$	$\dfrac{1}{2}$	$-\dfrac{\sqrt{3}}{2}$	$-\dfrac{\sqrt{3}}{3}$	$-\sqrt{3}$
$\dfrac{5\pi}{4}$	$-\dfrac{\sqrt{2}}{2}$	$-\dfrac{\sqrt{2}}{2}$	1	1	$-\dfrac{5\pi}{4}$	$\dfrac{\sqrt{2}}{2}$	$-\dfrac{\sqrt{2}}{2}$	-1	-1
$\dfrac{4\pi}{3}$	$-\dfrac{\sqrt{3}}{2}$	$-\dfrac{1}{2}$	$\sqrt{3}$	$\dfrac{\sqrt{3}}{3}$	$-\dfrac{4\pi}{3}$	$\dfrac{\sqrt{3}}{2}$	$-\dfrac{1}{2}$	$-\sqrt{3}$	$-\dfrac{\sqrt{3}}{3}$
$\dfrac{3\pi}{2}$	-1	0	$*$	0	$-\dfrac{3\pi}{2}$	1	0	$*$	0
$\dfrac{5\pi}{3}$	$-\dfrac{\sqrt{3}}{2}$	$\dfrac{1}{2}$	$-\sqrt{3}$	$-\dfrac{\sqrt{3}}{3}$	$-\dfrac{5\pi}{3}$	$\dfrac{\sqrt{3}}{2}$	$\dfrac{1}{2}$	$\sqrt{3}$	$\dfrac{\sqrt{3}}{3}$
$\dfrac{7\pi}{4}$	$-\dfrac{\sqrt{2}}{2}$	$\dfrac{\sqrt{2}}{2}$	-1	-1	$-\dfrac{7\pi}{4}$	$\dfrac{\sqrt{2}}{2}$	$\dfrac{\sqrt{2}}{2}$	1	1
$\dfrac{11\pi}{6}$	$-\dfrac{1}{2}$	$\dfrac{\sqrt{3}}{2}$	$-\dfrac{\sqrt{3}}{3}$	$-\sqrt{3}$	$-\dfrac{11\pi}{6}$	$\dfrac{1}{2}$	$\dfrac{\sqrt{3}}{2}$	$\dfrac{\sqrt{3}}{3}$	$\sqrt{3}$
2π	0	1	0	$*$	-2π	0	1	0	$*$

*Not defined.

Table 8.2
Partial Table of Values of tan(*t*) and cot(*t*)

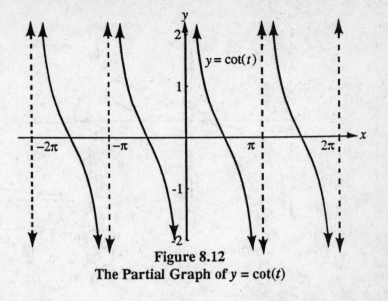

Figure 8.12
The Partial Graph of $y = \cot(t)$

Definition:
Consider the restricted tangent function with domain $(-\pi/2, \pi/2)$.
The **Inverse Tangent** function, denoted by $\tan^{-1}$, is defined by
$y = \tan^{-1}(x)$, if and only if $x = \tan(y)$. The symbol $\tan^{-1}(x)$
is read "the inverse tangent of x."

We now make some observations about the Inverse Tangent function.

- The **domain** of $\tan^{-1}$ is $(-\infty, +\infty)$, which is equal to the range of the restricted Tangent function.
- The **range** of $\tan^{-1}$ is $(-\pi/2, \pi/2)$, which is equal to the domain of the restricted Tangent function.
- The graph of $\tan^{-1}$ is obtained by reflecting the graph of tan restricted to $(-\pi/2, \pi/2)$ about the line with equation $y = x$. (The graphs of the restricted tangent function and $\tan^{-1}$ are given in Figure 8.13.)
- The $\tan^{-1}$ function is **one-to-one**.
- The $\tan^{-1}$ function has an **inverse as a function**. The inverse function is the tangent function restricted to $(-\pi/2, \pi/2)$.

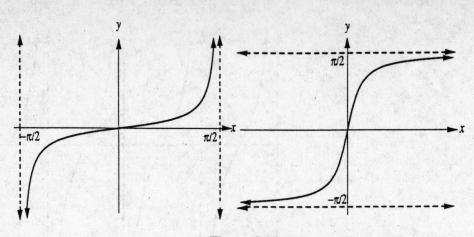

Figure 8.13

a) **The Graph of $= \tan(x)$ restricted to $(-\pi/2, \pi/2)$**

b) **The Graph of $y = \tan^{-1}(x)$**

EXERCISE 8.32

Evaluate $\tan^{-1}(-1)$.

SOLUTION 8.32

We are looking for a value of t such that $\tan(t) = -1$ and such that $-\pi/2 \leq t < \pi/2$. But,

$$\tan(-\pi/4) = -1 \quad \text{and} \quad -\pi/2 \leq -\pi/4 < \pi/2.$$

There are no other values of t satisfying these conditions. Therefore, $\tan^{-1}(-1) = -\pi/4$.

EXERCISE 8.33

Evaluate $\tan^{-1}(0)$.

SOLUTION 8.33

We are looking for a value of t such that $\tan(t) = 0$ and such that $-\pi/2 \leq t < \pi/2$. But,

$$\tan(0) = 0 \quad \text{and} \quad -\pi/2 \leq 0 < \pi/2.$$

There are no other values of t satisfying these conditions. Therefore, $\tan^{-1}(0) = 0$.

EXERCISE 8.34

Evaluate $\sin(\tan^{-1}(-1))$.

SOLUTION 8.34

Let $t = \tan^{-1}(-1)$. Then, $\tan(t) = -1$ such that $-\pi/2 \le t < \pi/2$. We determine that $t = -\pi/4$. Hence,

$$\sin(\tan^{-1}(-1)) = \sin(-\pi/4)$$
$$= \frac{-\sqrt{2}}{2}.$$

EXERCISE 8.35

Evaluate $\cos(2 \cdot \tan^{-1}(-1))$.

SOLUTION 8.35

Let $t = \tan^{-1}(1)$. Then, $\tan(t) = 1$ such that $-\pi/2 \le t < \pi/2$. We determine that $t = \pi/4$. Hence,

$$\cos(2 \cdot \tan^{-1}(-1)) = \cos(2(\pi/4))$$
$$= \cos(\pi/2)$$
$$= 0.$$

Definition:

Consider the restricted cotangent function with domain $(0, \pi)$. The **Inverse Cotangent** function, denoted by $\cot^{-1}$, is defined by $y = \cot^{-1}(x)$, if and only if $x = \cot(y)$. The symbol $\cot^{-1}(x)$ is read "the inverse cotangent of x."

We now make some observations about the Inverse Cotangent function.

- The **domain** of $\cot^{-1}$ is $(-\infty, +\infty)$, which is equal to the range of the restricted cotangent function.

- The **range** of $\cot^{-1}$ is $(0, \pi)$, which is equal to the domain of the restricted cotangent function.

- The graph of $\cot^{-1}$ is obtained by reflecting the graph of the cot function, restricted to $(0, \pi)$, about the line with equation $y = x$. (The graphs of the restricted cot function and the $\cot^{-1}$ function are given in Figure 8.14.)

- The $\cot^{-1}$ function is **one-to-one**.

- The $\cot^{-1}$ function has an **inverse as a function**. The inverse function is the cotangent function restricted to $(0, \pi)$.

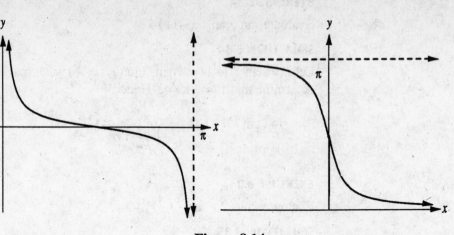

Figure 8.14

a) **The Graph of $y = \cot(x)$ restricted to $(0, \pi)$**

b) **The Graph of $y = \cot^{-1}(x)$**

EXERCISE 8.36

Evaluate $\cot^{-1}(\sqrt{3}/3)$.

SOLUTION 8.36

We are looking for a value of t such that $\cot(t) = \sqrt{3}/3$ and such that $0 < t < \pi$. But,

$$\cot(\pi/3) = \sqrt{3}/3 \qquad \text{and} \qquad 0 < \pi/3 < \pi.$$

There are no other values of t satisfying these conditions. Therefore, $\cot^{-1}(\sqrt{3}/3) = \pi/3$.

EXERCISE 8.37

Evaluate $\cot^{-1}(0)$.

SOLUTION 8.37

We are looking for a value of t such that $\cot(t) = 0$ and such that $0 < t < \pi$. But,

$$\cot(\pi/2) = 0 \qquad \text{and} \qquad 0 < \pi/2 < \pi.$$

There are no other values of t satisfying these conditions. Therefore, $\cot^{-1}(0) = \pi/2$.

EXERCISE 8.38

Evaluate $\cot(\cot^{-1}(-3))$.

SOLUTION 8.38

We have the composition of the functions cot and $\cot^{-1}$, which is the Identity function. Hence,

$$\cot(\cot^{-1}(-3)) = -3.$$

EXERCISE 8.39

Evaluate $\cos\left(\cot^{-1}(\sqrt{3}/3)\right)$.

SOLUTION 8.39

Let $t = \cot^{-1}(\sqrt{3}/3)$. Then, $\cot(t) = \sqrt{3}/3$ such that $0 < t < \pi$. We determine that $t = \pi/3$. Hence,

$$\cos\left(\cot^{-1}(\sqrt{3}/3)\right) = \cos(\pi/3)$$
$$= 1/2.$$

EXERCISE 8.40

Evaluate $-3 \cdot \sin\left(\cot^{-1}(-\sqrt{3})\right)$.

SOLUTION 8.40

Let $t = \cot^{-1}(-\sqrt{3})$. Then, $\cot(t) = -\sqrt{3}$ such that $0 < t < \pi$. We determine that $t = 5\pi/6$. Hence,

$$-3 \cdot \sin\left(\cot^{-1}(-\sqrt{3})\right) = -3 \cdot \sin(5\pi/6)$$
$$= -3(1/2)$$
$$= -3/2.$$

EXERCISE 8.41

Evaluate $-5 \cdot \cot\left(\tan^{-1}(1) + \sin^{-1}(\sqrt{2}/2)\right)$.

SOLUTION 8.41

Let $t = \tan^{-1}(1)$. Then, $\tan(t) = 1$ and $-\pi/2 < t < \pi/2$. We determine that $t = \pi/4$.

Let $p = \sin^{-1}(\sqrt{2}/2)$. Then, $\sin(p) = \sqrt{2}/2$ and $-\pi/2 < p < \pi/2$. We determine that $p = \pi/4$. Substituting for t and p, we have

$$-5 \cdot \cot\left(\tan^{-1}(1) + \sin^{-1}(\sqrt{2}/2)\right) = -5 \cdot \cot(\pi/4 + \pi/4)$$
$$= -5 \cdot \cot(\pi/2)$$
$$= -5(0)$$
$$= 0.$$

EXERCISE 8.42

Evaluate $\cos \left(2 \cdot \tan^{-1} (\sqrt{3}/3) - 4 \cdot \cot^{-1} (1) \right)$.

SOLUTION 8.42

Let $t = \tan^{-1} (\sqrt{3}/3)$. Then, $\tan (t) = \sqrt{3}/3$ and $-\pi/2 < t < \pi/2$. We determine that $t = \pi/6$.

Let $p = \cot^{-1} (1)$. Then, $\cot (p) = 1$ and $0 < p < \pi$. We determine that $p = \pi/4$. Substituting for t and p, we have

$$\cos \left(2 \cdot \tan^{-1} (\sqrt{3}/3) - 4 \cdot \cot^{-1} (1) \right) = \cos (2 (\pi/6) - 4 (\pi/4))$$
$$= \cos (\pi/3 - \pi)$$
$$= \cos (-2\pi/3)$$
$$= -1/2.$$

8.5 THE SECANT AND COSECANT FUNCTIONS AND THEIR INVERSES

There are two remaining circular functions. These functions are defined in terms of the reciprocals of the sine and cosine functions, whenever the reciprocals exist.

Definition:

Consider the cosine function defined by $x = \cos (t) =$ the x-coordinate of $W (t)$. The **Secant** function, denoted by sec, is the reciprocal of the cosine function; that is,

$$\sec (t) = \frac{1}{\cos (t)} \text{ for all real values of } t \text{ for which } \cos (t) \neq 0.$$

If $\cos (t) = x$ for all real values of t, we determine that $\sec (t) = \dfrac{1}{x}$, provided that $x \neq 0$. We now make some observations about the secant function.

- Since $x = \cos (t) = 0$ when t is an odd integer multiple of $\pi/2$, we determine that the **domain** of sec is the set of all **real numbers** that are **not odd integer multiples of $\pi/2$**.

- Since the range of cos is $[-1, 1]$, the range of sec is the set of the reciprocal values in the interval. Hence, the **range** of sec is $(-\infty, -1] \cup [1, +\infty)$. (This will be substantiated, in part, by an examination of the graph of sec.)

t	$W(t)$	$\sec(t)$	$\csc(t)$	t	$W(t)$	$\sec(t)$	$\csc(t)$
0	$(1, 0)$	1	$*$				
$\dfrac{\pi}{6}$	$\left(\dfrac{\sqrt{3}}{2}, \dfrac{1}{2}\right)$	$\dfrac{2\sqrt{3}}{3}$	2	$-\dfrac{\pi}{6}$	$\left(\dfrac{\sqrt{3}}{2}, -\dfrac{1}{2}\right)$	$\dfrac{2\sqrt{3}}{3}$	-2
$\dfrac{\pi}{4}$	$\left(\dfrac{\sqrt{2}}{2}, \dfrac{\sqrt{2}}{2}\right)$	$\sqrt{2}$	$\sqrt{2}$	$-\dfrac{\pi}{4}$	$\left(\dfrac{\sqrt{2}}{2}, -\dfrac{\sqrt{2}}{2}\right)$	$\sqrt{2}$	$-\sqrt{2}$
$\dfrac{\pi}{3}$	$\left(\dfrac{1}{2}, \dfrac{\sqrt{3}}{2}\right)$	2	$\dfrac{2\sqrt{3}}{3}$	$-\dfrac{\pi}{3}$	$\left(\dfrac{1}{2}, -\dfrac{\sqrt{3}}{2}\right)$	2	$-\dfrac{2\sqrt{3}}{3}$
$\dfrac{\pi}{2}$	$(0, 1)$	$*$	1	$-\dfrac{\pi}{2}$	$(0, -1)$	$*$	-1
$\dfrac{2\pi}{3}$	$\left(-\dfrac{1}{2}, \dfrac{\sqrt{3}}{2}\right)$	-2	$\dfrac{2\sqrt{3}}{3}$	$-\dfrac{2\pi}{3}$	$\left(-\dfrac{1}{2}, -\dfrac{\sqrt{3}}{2}\right)$	-2	$-\dfrac{2\sqrt{3}}{3}$
$\dfrac{3\pi}{4}$	$\left(-\dfrac{\sqrt{2}}{2}, \dfrac{\sqrt{2}}{2}\right)$	$-\sqrt{2}$	$\sqrt{2}$	$-\dfrac{3\pi}{4}$	$\left(-\dfrac{\sqrt{2}}{2}, -\dfrac{\sqrt{2}}{2}\right)$	$-\sqrt{2}$	$-\sqrt{2}$
$\dfrac{5\pi}{6}$	$\left(-\dfrac{\sqrt{3}}{2}, \dfrac{1}{2}\right)$	$-\dfrac{2\sqrt{3}}{3}$	2	$-\dfrac{5\pi}{6}$	$\left(-\dfrac{\sqrt{3}}{2}, -\dfrac{1}{2}\right)$	$-\dfrac{2\sqrt{3}}{3}$	-2
π	$(-1, 0)$	-1	$*$	$-\pi$	$(-1, 0)$	-1	$*$
$\dfrac{7\pi}{6}$	$\left(-\dfrac{\sqrt{3}}{2}, -\dfrac{1}{2}\right)$	$-\dfrac{2\sqrt{3}}{3}$	-2	$-\dfrac{7\pi}{6}$	$\left(-\dfrac{\sqrt{3}}{2}, \dfrac{1}{2}\right)$	$-\dfrac{2\sqrt{3}}{3}$	2
$\dfrac{5\pi}{4}$	$\left(-\dfrac{\sqrt{2}}{2}, -\dfrac{\sqrt{2}}{2}\right)$	$-\sqrt{2}$	$-\sqrt{2}$	$-\dfrac{5\pi}{4}$	$\left(-\dfrac{\sqrt{2}}{2}, \dfrac{\sqrt{2}}{2}\right)$	$-\sqrt{2}$	$\sqrt{2}$
$\dfrac{4\pi}{3}$	$\left(-\dfrac{1}{2}, -\dfrac{\sqrt{3}}{2}\right)$	-2	$-\dfrac{2\sqrt{3}}{3}$	$-\dfrac{4\pi}{3}$	$\left(-\dfrac{1}{2}, \dfrac{\sqrt{3}}{2}\right)$	-2	$\dfrac{2\sqrt{3}}{3}$
$\dfrac{3\pi}{2}$	$(0, -1)$	$*$	-1	$-\dfrac{3\pi}{2}$	$(0, 1)$	$*$	1
$\dfrac{5\pi}{3}$	$\left(\dfrac{1}{2}, -\dfrac{\sqrt{3}}{2}\right)$	2	$-\dfrac{2\sqrt{3}}{3}$	$-\dfrac{5\pi}{3}$	$\left(\dfrac{1}{2}, \dfrac{\sqrt{3}}{2}\right)$	2	$\dfrac{2\sqrt{3}}{3}$
$\dfrac{7\pi}{4}$	$\left(\dfrac{\sqrt{2}}{2}, -\dfrac{\sqrt{2}}{2}\right)$	$\sqrt{2}$	$-\sqrt{2}$	$-\dfrac{7\pi}{4}$	$\left(\dfrac{\sqrt{2}}{2}, \dfrac{\sqrt{2}}{2}\right)$	$\sqrt{2}$	$\sqrt{2}$
$\dfrac{11\pi}{6}$	$\left(\dfrac{\sqrt{3}}{2}, -\dfrac{1}{2}\right)$	$\dfrac{2\sqrt{3}}{3}$	-2	$-\dfrac{11\pi}{6}$	$\left(\dfrac{\sqrt{3}}{2}, \dfrac{1}{2}\right)$	$\dfrac{2\sqrt{3}}{3}$	2
2π	$(1, 0)$	1	$*$	-2π	$(1, 0)$	1	$*$

*Not defined.

Table 8.3
Partial Table of Values for sec(t) and csc(t)

- To graph the secant function, we use the reciprocals of the cos (t) values given in Table 8.1. These reciprocal values are given in Table 8.3, and a partial graph of the secant function is given in Figure 8.15.
- The secant function is **not one-to-one**.
- The secant function does **not** have an **inverse as a function**.
- The **period** of the secant function is 2π.

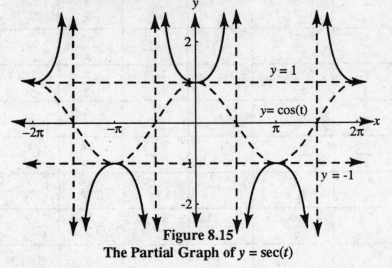

Figure 8.15
The Partial Graph of $y = \sec(t)$

Definition:
Consider the sine function defined by $y = \sin(t) =$ the y-coordinate of $W(t)$. The **Cosecant** function, denoted by csc, is the reciprocal of the sine function; that is,

$$\csc(t) = \frac{1}{\sin(t)} \text{ for all real values of } t, \text{ provided that } \sin(t) \neq 0.$$

If $\sin(t) = y$ for all real values of t, we determine that $\csc(t) = \frac{1}{y}$, provided that $y \neq 0$. We now make some observations about the cosecant function.

- Since $y = \sin(t) = 0$ when t is an integer multiple of π, we determine that the **domain** of csc is the set of all **real numbers** that are **not integer multiples of** π.
- Since the range of cos is $[-1, 1]$, the range of csc is the set of the reciprocal values in the interval. Hence, the **range** of csc is $(-\infty, -1] \cup [1, +\infty)$. (Again, this will be substantiated, in part, by an examination of the graph of csc.)
- To graph the cosecant function, we use the reciprocals of the sin (t) values given in Table 8.1. These reciprocal values are given in Table

8.3, and a partial graph of the Cosecant function is given in Figure 8.16.

- The cosecant function is **not one-to-one**.
- The cosecant function does **not** have an **inverse as a function.**
- The **period** of the cosecant function is **2π.**

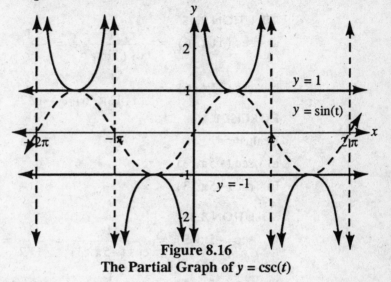

Figure 8.16
The Partial Graph of $y = \csc(t)$

EXERCISE 8.43

Evaluate

a) $\sec(-\pi/4)$

b) $\csc(-\pi/4)$

SOLUTION 8.43

a) $\sec(-\pi/4) = \dfrac{1}{\cos(-\pi/4)} = \dfrac{1}{\sqrt{2}/2} = \sqrt{2}$

b) $\csc(-\pi/4) = \dfrac{1}{\sin(-\pi/4)} = \dfrac{1}{-\sqrt{2}/2} = -\sqrt{2}$

EXERCISE 8.44

Evaluate

a) $\sec(2\pi/3)$

b) $\csc(2\pi/3)$

SOLUTION 8.44

a) $\sec(2\pi/3) = \dfrac{1}{\cos(2\pi/3)} = \dfrac{1}{-1/2} = -2$

b) $\csc(2\pi/3) = \dfrac{1}{\sin(2\pi/3)} = \dfrac{1}{\sqrt{3}/2} = \dfrac{2\sqrt{3}}{3}$

EXERCISE 8.45

Evaluate

a) $\sec(11\pi/6)$

b) $\csc(11\pi/6)$

SOLUTION 8.45

a) $\sec(11\pi/6) = \dfrac{1}{\cos(11\pi/6)} = \dfrac{1}{\sqrt{3}/2} = \dfrac{2\sqrt{3}}{3}$

b) $\csc(11\pi/6) = \dfrac{1}{\sin(11\pi/6)} = \dfrac{1}{-1/2} = -2$

EXERCISE 8.46

Evaluate

a) $\sec(-5\pi/3)$

b) $\csc(-5\pi/3)$

SOLUTION 8.46

a) $\sec(-5\pi/3) = \dfrac{1}{\cos(-5\pi/3)} = \dfrac{1}{1/2} = 2$

b) $\csc(-5\pi/3) = \dfrac{1}{\sin(-5\pi/3)} = \dfrac{1}{\sqrt{3}/2} = \dfrac{2\sqrt{3}}{3}$

EXERCISE 8.47

Evaluate $\sec(t)$ and $\csc(t)$ for the indicated $W(t)$

a) $W(t) = \left(\dfrac{1}{4}, \dfrac{\sqrt{15}}{4}\right)$

b) $W(t) = \left(\dfrac{-1}{3}, \dfrac{2\sqrt{2}}{3}\right)$

c) $W(t) = \left(\dfrac{-\sqrt{5}}{5}, \dfrac{-2\sqrt{5}}{5}\right)$

d) $W(t) = \left(\dfrac{1}{8}, \dfrac{-\sqrt{63}}{8}\right)$

SOLUTION 8.47

a)   Since $W(t) = \left(\dfrac{1}{4}, \dfrac{\sqrt{15}}{4}\right)$, $\cos(t) = \dfrac{1}{4}$ and $\sin(t) = \dfrac{\sqrt{15}}{4}$.
Hence,

$$\sec(t) = \frac{1}{\cos(t)} = \frac{1}{1/4} = 4$$

and

$$\csc(t) = \frac{1}{\sin(t)} = \frac{1}{\sqrt{15}/4} = \frac{4\sqrt{15}}{15}$$

b) Since $W(t) = \left(\dfrac{-1}{3}, \dfrac{2\sqrt{2}}{3}\right)$, $\cos(t) = \dfrac{-1}{3}$ and $\sin(t) = \dfrac{2\sqrt{2}}{3}$.
Hence,

$$\sec(t) = \frac{1}{\cos(t)} = \frac{1}{-1/3} = -3$$

and

$$\csc(t) = \frac{1}{\sin(t)} = \frac{1}{2\sqrt{2}/3} = \frac{3\sqrt{2}}{4}$$

c) Since $W(t) = \left(\dfrac{-\sqrt{5}}{5}, \dfrac{-2\sqrt{5}}{5}\right)$, $\cos(t) = \dfrac{-\sqrt{5}}{5}$ and $\sin(t) = \dfrac{-2\sqrt{5}}{5}$.
Hence,

$$\sec(t) = \frac{1}{\cos(t)} = \frac{1}{-\sqrt{5}/5} = -\sqrt{5}$$

and

$$\csc(t) = \frac{1}{\sin(t)} = \frac{1}{-2\sqrt{5}/5} = \frac{-\sqrt{5}}{2}$$

d) Since $W(t) = \left(\dfrac{1}{8}, \dfrac{-\sqrt{63}}{8}\right)$, $\cos(t) = \dfrac{1}{8}$ and $\sin(t) = \dfrac{-\sqrt{63}}{8}$.
Hence,

$$\sec(t) = \frac{1}{\cos(t)} = \frac{1}{1/8} = 8$$

and

$$\csc(t) = \frac{1}{\sin(t)} = \frac{1}{-\sqrt{63}/8} = \frac{-8\sqrt{63}}{63}$$

Inverse Secant and Inverse Cosecant Functions

Definition:
Consider the restricted Secant function with domain $[0, \pi/2) \cup (\pi/2, \pi]$. The **Inverse Secant** function, denoted by $\sec^{-1}$, is defined by $y = \sec^{-1}(x)$, if and only if $x = \sec(y)$. The symbol $\sec^{-1}(x)$ is read "the inverse secant of x."

We now make some observations about the Inverse Secant function.
- The **domain** of $\sec^{-1}$ is $(-\infty, -1] \cup [1, +\infty)$, which is equal to the range of the restricted sec function.
- The **range** of $\sec^{-1}$ is $[0, \pi/2) \cup (\pi/2, \pi]$, which is equal to the domain of the restricted sec function.
- The graph of $\sec^{-1}$ is obtained by reflecting the graph of sec, restricted to $[0, \pi/2) \cup (\pi/2, \pi]$, about the line with equation $y = x$. (The graphs of the restricted sec function and $\sec^{-1}$ are given in Figure 8.17.)

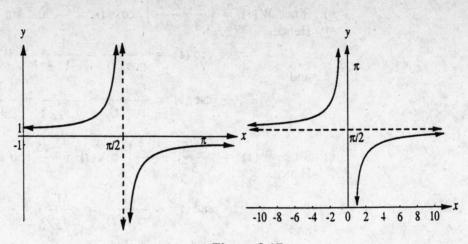

Figure 8.17

a) **The Graph of** $y = \sec(x)$ b) **The Graph of** $y = \sec^{-1}(x)$
    restricted to $[0, \pi/2) \cup (\pi/2, \pi]$

- The $\sec^{-1}$ function is **one-to-one**.

- The $\sec^{-1}$ function has an **inverse as a function**. It is the restricted secant function.

EXERCISE 8.48

Evaluate $\sec^{-1}(-\sqrt{2})$.

SOLUTION 8.48

We are looking for a value of t such that $\sec(t) = -\sqrt{2}$ and such that t is in the set $[0, \pi/2) \cup (\pi/2, \pi]$. But, $t = 3\pi/4$ satisfies these conditions. There are no other values of t that satisfy these conditions. Hence, $\sec^{-1}(-\sqrt{2}) = 3\pi/4$.

EXERCISE 8.49

Evaluate $\sec^{-1}(-2)$.

SOLUTION 8.49

We are looking for a value of t such that $\sec(t) = -2$ and such that t is in the set $[0, \pi/2) \cup (\pi/2, \pi]$. But, $t = 2\pi/3$ satisfies these conditions. There are no other values of t that satisfy these conditions. Hence, $\sec^{-1}(-2) = 2\pi/3$.

EXERCISE 8.50

Evaluate $\sec^{-1}(-1)$.

SOLUTION 8.50

We are looking for a value of t such that $\sec(t) = -1$ and such that t is in the set $[0, \pi/2) \cup (\pi/2, \pi]$. But, $t = \pi$ satisfies these conditions. There are no other values of t that satisfy these conditions. Hence, $\sec^{-1}(-1) = \pi$.

EXERCISE 8.51

Evaluate $\sec^{-1}(-2\sqrt{3}/3)$.

SOLUTION 8.51

We are looking for a value of t such that $\sec(t) = -2\sqrt{3}/3$ and such that t is in the set $[0, \pi/2) \cup (\pi/2, \pi]$. But, $t = 5\pi/6$ satisfies these conditions. There are no other values of t that satisfy these conditions. Hence, $\sec^{-1}(-2\sqrt{3}/3) = 5\pi/6$.

Definition:

Consider the restricted cosecant function with domain $[-\pi/2, 0) \cup (0, \pi/2]$. The **Inverse Cosecant** function, denoted by $\csc^{-1}$, is defined by $y = \csc^{-1}(x)$, if and only if $x = \csc(y)$. The symbol $\csc^{-1}(x)$ is read "the inverse cosecant of x."

We now make some observations about the Inverse Cosecant function.

- The **domain** of $\csc^{-1}$ is $(-\infty, -1] \cup [1, +\infty)$, which is equal to the range of the restricted cosecant function.
- The **range** of $\csc^{-1}$ is $[-\pi/2) \cup (0, \pi/2]$, which is equal to the domain of the restricted cosecant function.
- The graph of $\csc^{-1}$ is obtained by reflecting the graph of csc, restricted to $[-\pi/2, 0) \cup (0, \pi/2]$, about the line with equation $y = x$. (The graphs of the restricted cosecant function and $\csc^{-1}$ are given in Figure 8.18.)
- The $\csc^{-1}$ function is **one-to-one**.
- The $\csc^{-1}$ function has an **inverse as a function**. It is the restricted cosecant function.

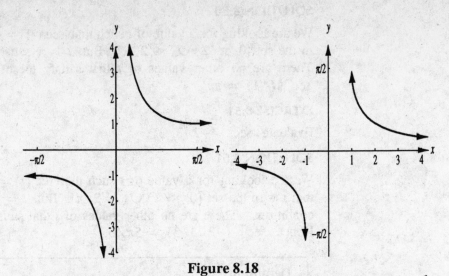

Figure 8.18

a) **The Graph of** $y = \csc(x)$
   restricted to $[-\pi/2, 0) \cup (0, \pi/2]$

b) **The Graph of** $y = \csc^{-1}(x)$

EXERCISE 8.52

Evaluate $\csc^{-1}(2)$.

SOLUTION 8.52

We are looking for a value of t such that $\csc(t) = 2$ and such that t is in
the set $[-\pi/2, 0) \cup (0, \pi/2]$. But, $t = \pi/6$ satisfies these conditions.
There are no other values of t that satisfy these conditions. Hence,

$$\csc^{-1}(2) = \pi/6.$$

EXERCISE 8.53

Evaluate $\csc^{-1}(-1)$.

SOLUTION 8.53

We are looking for a value of t such that $\csc(t) = -1$ and such that t is
in the set $[-\pi/2, 0) \cup (0, \pi/2]$. But, $t = -\pi/2$ satisfies these conditions. There are no other values of t that satisfy these conditions. Hence,
$$\csc^{-1}(-1) = -\pi/2.$$

EXERCISE 8.54

Evaluate $\csc^{-1}(-\sqrt{2})$.

SOLUTION 8.54

We are looking for a value of t such that $\csc(t) = -\sqrt{2}$ and such that t is
in the set $[-\pi/2, 0) \cup (0, \pi/2]$. But, $t = -\pi/4$ satisfies these condi-

tions. There are no other values of t that satisfy these conditions. Hence,

$$\csc^{-1}(-\sqrt{2}) = -\pi/4.$$

EXERCISE 8.55

Evaluate $3 \cdot \sin\left(2 \cdot \csc^{-1}(-\sqrt{2})\right)$.

SOLUTION 8.55

Let $t = \csc^{-1}(-\sqrt{2})$. Then, from the previous exercise, $t = -\pi/4$. Hence,

$$\begin{aligned}
3 \cdot \sin\left(2 \cdot \csc^{-1}(-\sqrt{2})\right) &= 3 \cdot \sin(2(-\pi/4)) \\
&= 3 \cdot \sin(-\pi/2) \\
&= 3 \cdot (-1) \\
&= -3.
\end{aligned}$$

EXERCISE 8.56

Evaluate $-4 \cdot \cos\left(2 \cdot \sec^{-1}(-2/\sqrt{3})\right)$.

SOLUTION 8.56

Let $t = \sec^{-1}(-2/\sqrt{3})$. We determine that $t = 5\pi/6$. Hence,

$$\begin{aligned}
-4 \cdot \cos\left(2 \cdot \sec^{-1}(-2/\sqrt{3})\right) &= -(4 \cdot \cos(2(5\pi/6))) \\
&= -4 \cdot \cos(5\pi/3) \\
&= -4(1/2) \\
&= -2.
\end{aligned}$$

EXERCISE 8.57

Evaluate $\sec\left(2 \cdot \csc^{-1}(-\sqrt{2}) - 3 \cdot \tan^{-1}(-1)\right)$.

SOLUTION 8.57

Let $t = \csc^{-1}(-2)$. We determine that $t = -\pi/4$.
Let $p = \tan^{-1}(-1)$. We determine that $p = -\pi/4$. Hence,

$$\begin{aligned}
\sec\left(2 \cdot \csc^{-1}(-\sqrt{2}) - 3 \cdot \tan^{-1}(-1)\right) &= \sec(2(-\pi/4) - 3(-\pi/4)) \\
&= \sec(-\pi/2 + 3\pi/4) \\
&= \sec(\pi/4) \\
&= \sqrt{2}.
\end{aligned}$$

8.6 FUNDAMENTAL IDENTITIES

Certain basic relations may be established among circular functions. Such relations are referred to as the fundamental or elementary identities of trigonometry. The most elementary of these relations is associated with the unit circle satisfying the equation

$$x^2 + y^2 = 1$$

such that

$$x = \cos(t) \qquad \text{and} \qquad y = \sin(t).$$

By simple substitution, we get the basic identity

$$\cos^2(t) + \sin^2(t) = 1$$

which is valid for all real values of t. The symbol $\sin^2(t)$ is used instead of $(\sin(t))^2$. In general, the symbol $\sin^n(t)$ is used instead of $(\sin(t))^n$, provided that n is a *positive integer*. The notation is *not* used if n is a negative integer. The symbol $(\sin(t))^{-1}$ does, however, mean $\frac{1}{\sin(t)}$, provided that $\sin(t) \neq 0$. Similar comments hold for the other circular functions. You should be very careful to use the symbols correctly.

From the definitions for tan, cot, sec, and csc, we readily obtain the following relations.

$$\tan(t) = \frac{\sin(t)}{\cos(t)}, \text{ provided that } \cos(t) \neq 0$$

$$\cot(t) = \frac{\cos(t)}{\sin(t)}, \text{ provided that } \sin(t) \neq 0$$

$$\sec(t) = \frac{1}{\cos(t)}, \text{ provided that } \cos(t) \neq 0$$

$$\csc(t) = \frac{1}{\sin(t)}, \text{ provided that } \sin(t) \neq 0$$

$$\cot(t) = \frac{1}{\tan(t)}, \text{ provided that } \tan(t) \neq 0$$

From the relation $\cos^2(t) + \sin^2(t) = 1$, we obtain an additional relation by dividing both sides of the equation by $\cos^2(t)$, provided that $\cos^2(t) \neq 0$. Hence,

$$\frac{\cos^2(t)}{\cos^2(t)} + \frac{\sin^2(t)}{\cos^2(t)} = \frac{1}{\cos^2(t)}$$

$$1 + \frac{\sin^2(t)}{\cos^2(t)} = \frac{1}{\cos^2(t)}$$

or

$$\boxed{1 + \tan^2(t) = \sec^2(t), \text{ provided that } \cos(t) \neq 0}$$

In a similar manner, if we divide both sides of $\sin^2(t) + \cos^2(t) = 1$ by $\sin^2(t)$, provided that $\sin^2(t) \neq 0$, we obtain the following identity.

$$\boxed{1 + \cot^2(t) = \csc^2(t), \text{ provided that } \sin(t) \neq 0}$$

These eight basic relations are called **identities** in the sense that they are true for all real values of t for which the respective circular functions are defined. Identities are useful when attempting to simplify trigonometric expressions that may be used in applied problems.

EXERCISE 8.58

Show that $\tan(t)(\sin(t) + \cot(t)\cos(t)) = \sec(t)$.

SOLUTION 8.58

Since the left-hand side of the given equation is more complex than the right-hand side, we will attempt to rewrite the left-hand side expression to match the expression on the right-hand side exactly, without going across the equal sign.

$\tan(t)(\sin(t) + \cot(t)\cos(t))$

$= \tan(t)\sin(t) + \tan(t)\cot(t)\cdot\cos(t)$ Expanding

$= \tan(t)\sin(t) + \tan(t)\dfrac{1}{\tan(t)}\cos(t)$ Substituting

$= \tan(t)\sin(t) + \cos(t)$ Simplifying, provided that $\tan(t) \neq 0$

$= \dfrac{\sin(t)}{\cos(t)}\sin(t) + \cos(t)$ Substituting

$$= \frac{\sin^2(t)}{\cos(t)} + \cos(t) \qquad \text{Multiplying}$$

$$= \frac{\sin^2(t) + \cos^2(t)}{\cos(t)} \qquad \text{Combining}$$

$$= \frac{1}{\cos(t)} \qquad \text{Substituting}$$

$$= \sec(t)$$

Hence, we have established that the given equation is true for all values of t for which all of the functions involved are defined and for which $\tan(t) \neq 0$.

EXERCISE 8.59

Show that $\dfrac{\sec(t)}{\sin(t)} - \dfrac{\sin(t)}{\cos(t)} = \cot(t)$.

SOLUTION 8.59

Again, we will attempt to rewrite the left-hand side expression to match the right-hand side expression exactly.

$$\frac{\sec(t)}{\sin(t)} - \frac{\sin(t)}{\cos(t)}$$

$$= \frac{\sec(t)\cos(t) - \sin^2(t)}{\sin(t)\cos(t)} \qquad \text{Combining}$$

$$= \frac{\dfrac{1}{\cos(t)}\cos(t) - \sin^2(t)}{\sin(t)\cos(t)} \qquad \text{Substituting}$$

$$= \frac{1 - \sin^2(t)}{\sin(t)\cos(t)} \qquad \text{Simplifying, provided that } \cos(t) \neq 0$$

$$= \frac{\cos^2(t)}{\sin(t)\cos(t)} \qquad \text{Substituting}$$

$$= \frac{\cos(t)}{\sin(t)} \qquad \text{Dividing, provided that } \cos(t) \neq 0$$

$$= \cot(t)$$

Hence, we have established that the above equation is true for all values of t for which all of the functions involved are defined and for which $\cos(t) \neq 0$.

In this chapter, we introduced the Wrapping function, W, that maps the set of real numbers onto the unit circle, C. Using the coordinates of an arbitrary point, W(t), on the circumference of C, we defined and discussed the circular functions. Values of circular functions were obtained using geometric reasoning and tables. By restricting the domains of each of the circular functions, we introduced and discussed the inverse function for each of them. From the definitions of the circular functions, we established some fundamental identities.

SUPPLEMENTAL EXERCISES

Identify each of the following statements as being true or false and give a reason for each of your answers.

1. There exists a value of t such that $\sin(t) = 2\sqrt{2}$.

2. There exists a value of t such that $\cot(t)$ is not defined.

3. There exists a value of t such that $\sec(t) = -0.8$.

4. The range of $\sec^{-1}$ is the set of all positive real numbers that are greater than or equal to 1.

5. The domain of cot is the set of all real numbers that are not multiples of π.

6. The cosine function is an increasing function on $[0, \pi/2]$.

7. $1 + \tan^2(7\pi/8) = \sec^2(-7\pi/8)$

8. $1 + \cot^2(t) = \csc^2(t)$ if and only if $0 < t < \pi$.

9. The secant function is periodic, with a period of π.

10. $\sin(0) = \sin(\pi)$

11. The domain of the sine function is equal to the range of the tangent function.

12. $W(t) = W(t - 2\pi)$

13. $W(\pi) = W(-5\pi)$

14. The smallest positive value of k for which $\tan(t + k) = \tan(t)$ is 2π.

15. The graph of the Cosecant function is not contained between the lines $y = -1/2$ and $y = 1/2$.

16. The Tangent function is increasing on the interval $(-\pi/2, \pi/2)$.

17. The Wrapping function, W, is one-to-one and onto C.

18. The cosine function is defined in terms of the y-coordinate of $W(t)$.

19. The tangent and cotangent functions are reciprocal functions.

20. The exact value for $\cos(\pi/4)$ is given as $\sqrt{2}/2$ and not 0.7071.

ANSWERS TO SUPPLEMENTAL EXERCISES

1. False; the value of $\sin(t)$ cannot exceed 1.

2. True; the cot function is not defined for all integer multiples of π.

3. False; there are no negative values of $\sec(t)$ that are greater than -1.

4. False; the range of $\sec^{-1}$ is the set $[0, \pi/2) \cup (\pi/2, \pi]$.

5. True.

6. False; on the given interval, the cosine function is a decreasing function.

7. True; $1 + \tan^2(7\pi/8) = \sec^2(7\pi/8)$, $\sec^2(-7\pi/8) = \sec^2(7\pi/8)$

8. False; the equation is true provided that t is not an integer multiple of π.

9. False; the secant function has a period of 2π.

10. True; $\sin(0) = \sin(\pi) = 0$.

11. True; they are both equal to $(-\infty, +\infty)$.

12. True; $W(t)$ and $W(t-2\pi)$ are different names for the same point on the unit circle.

13. True; $W(\pi) = W(-5\pi) = (-1, 0)$.

14. False; the smallest positive value of k is π.

15. True.

16. True.

17. False; the Wrapping function is onto C but it is not one-to-one.

18. False; the cosine function is defined in terms of the x-coordinate of $W(t)$.

19. True.

20. True; 0.7071 is an approximate value for $\cos(\pi/4)$.

9

Trigonometric Functions: Extensions of the Circular Functions

*T*he discussion of the circular functions will serve as the groundwork for considering applications of the trigonometric functions. However, before proceeding with these applications, we first direct our attention to some methods of computing more values for the circular functions. We ask, for instance, what relationships, if any, exist among the expressions $\cos(u)$, $\cos(v)$, and $\cos(u-v)$. We attempt to answer this question by introducing the addition formulas.

9.1 ADDITION FORMULAS

If F is an arbitrary function, then an addition formula relates function values such as $F(u)$, $F(v)$, and $F(u \pm v)$. In general, we have no reason to expect that $F(u \pm v)$ will be equal to $F(u) \pm F(v)$. For example, if $u = \pi/2$ and $v = \pi/3$, then $\cos(\pi/2 - \pi/3) = \cos(\pi/6) = \sqrt{3}/2$. However, $\cos(\pi/2) = 0$, $\cos(\pi/3) = 1/2$, and $\cos(\pi/2) - \cos(\pi/3) = 0 - 1/2 = -1/2$. Hence, $\cos(u-v) \neq \cos(u) - \cos(v)$. We shall start with any two real numbers, u and v, and consider $W(u)$ and $W(v)$. For illustration purposes only, let $u > v$. (See Figure 9.1.) Hence, $u - v > 0$ and, starting from $W(0)$ and proceeding along the circumference of C in a positive direction, we can locate $W(u-v)$ as indicated. It follows that the arc from $W(v)$ to $W(u)$ is equal in length to the arc from $W(0)$ to $W(u-v)$. Therefore, since congruent arcs subtend congruent chords, we have

$$\overline{W(u)\,W(v)} = \overline{W(0)\,W(u-v)}.$$

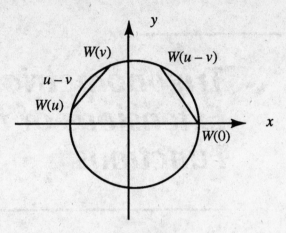

Figure 9.1

Using the undirected distance formula and noting that the coordinates of $W(0)$, $W(u)$, $W(v)$, and $W(u-v)$ are, respectively, $(1,0)$, $(\cos(u), \sin(u))$, $(\cos(v), \sin(v))$, and $(\cos(u-v), \sin(u-v))$, we have

$$\sqrt{(\cos(v) - \cos(u))^2 + (\sin(v) - \sin(u))^2}$$
$$= \sqrt{(\cos(u-v) - 1)^2 + (\sin(u-v) - 0)^2}.$$

Squaring both sides of this equation, we obtain

$$(\cos(v) - \cos(u))^2 + (\sin(v) - \sin(u))^2$$
$$= (\cos(u-v) - 1)^2 + (\sin(u-v) - 0)^2.$$

Expanding, we have

$$\cos^2(v) - 2\cdot\cos(u)\cos(v) + \cos^2(u) + \sin^2(v)$$
$$- 2\cdot\sin(u)\sin(v) + \sin^2(u)$$
$$= \cos^2(u-v) - 2\cdot\cos(u-v) + 1 + \sin^2(u-v).$$

But,
$$\cos^2(u) + \sin^2(u) = \cos^2(v) + \sin^2(v) = \cos^2(u-v) + \sin^2(u-v)$$
$$= 1.$$

Therefore, the equation simplifies to
$$2 - 2 \cdot \cos(u) \cos(v) - 2 \cdot \sin(u) \sin(v) = 2 - 2 \cdot \cos(u - v)$$
or
$$-2 \cdot \cos(u) \cos(v) - 2 \cdot \sin(u) \sin(v) = -2 \cdot \cos(u - v),$$
from which we obtain

$$\cos(u - v) = \cos(u) \cos(v) + \sin(u) \sin(v) \qquad (1)$$

Equation (1) is an identity that is true for all real values of u and v. Further addition formulas may be obtained by assigning specific values to u and v.

EXERCISE 9.1

Determine the value of $\cos(\pi/6)$ using circular function values of $\pi/2$ and $\pi/3$.

SOLUTION 9.1

Since $\pi/6 = \pi/2 - \pi/3$, we have
$$\begin{aligned}
\cos(\pi/6) &= \cos(\pi/2 - \pi/3) \\
&= \cos(\pi/2) \cos(\pi/3) + \sin(\pi/2) \sin(\pi/3) \\
&= (0)(1/2) + (1)(\sqrt{3}/2) \\
&= 0 + \sqrt{3}/2 \\
&= \sqrt{3}/2.
\end{aligned}$$

EXERCISE 9.2

Determine the value of $\cos(\pi/12)$.

SOLUTION 9.2

Since $\pi/12 = \pi/3 - \pi/4$, we have
$$\begin{aligned}
\cos(\pi/12) &= \cos(\pi/3 - \pi/4) \\
&= \cos(\pi/3) \cos(\pi/4) + \sin(\pi/3) \sin(\pi/4) \\
&= (1/2)(\sqrt{2}/2) + (\sqrt{3}/2)(\sqrt{2}/2) \\
&= \sqrt{2}/4 + \sqrt{6}/4 \\
&= \frac{\sqrt{2} + \sqrt{6}}{4}.
\end{aligned}$$

EXERCISE 9.3

Determine the value of $\cos(5\pi/12)$.

SOLUTION 9.3

Since $5\pi/12 = 2\pi/3 - \pi/4$, we have

$$\begin{aligned}
\cos(5\pi/12) &= \cos(2\pi/3 - \pi/4)\\
&= \cos(2\pi/3)\cos(\pi/4) + \sin(2\pi/3)\sin(\pi/4)\\
&= (-1/2)(\sqrt{2}/2) + (\sqrt{3}/2)(\sqrt{2}/2)\\
&= -\sqrt{2}/4 + \sqrt{6}/4\\
&= \frac{\sqrt{6} - \sqrt{2}}{4}.
\end{aligned}$$

Since $\cos(u - v) = \cos(u)\cos(v) + \sin(u)\sin(v)$ is true for all real values of u and v, it is true for $u = \pi/2$. Hence, we have

$$\begin{aligned}
\cos(\pi/2 - v) &= \cos(\pi/2)\cos(v) + \sin(\pi/2)\sin(v)\\
&= 0 \cdot \cos(v) + 1 \cdot \sin(v)\\
&= 0 + \sin(v)\\
&= \sin(v).
\end{aligned}$$

Therefore, we have

$$\boxed{\cos(\pi/2 - v) = \sin(v) \qquad\qquad (2)}$$

which is true for all real values of v.

EXERCISE 9.4

Rewrite each of the following in equivalent form using the cosine function.

a) $\sin(\pi/3)$

b) $\sin(7\pi/12)$

c) $\sin(\pi)$

d) $\sin(5\pi/4)$

e) $\sin(11\pi/6)$

SOLUTION 9.4

a) $\begin{aligned}[t]\sin(\pi/3) &= \cos(\pi/2 - \pi/3)\\ &= \cos(\pi/6)\end{aligned}$

b) $\begin{aligned}[t]\sin(7\pi/12) &= \cos(\pi/2 - 7\pi/12)\\ &= \cos(-\pi/12)\end{aligned}$

c) $\sin(\pi) = \cos(\pi/2 - \pi)$

$\qquad\qquad = \cos(-\pi/2)$

d) $\sin(5\pi/4) = \cos(\pi/2 - 5\pi/4)$

$\qquad\qquad\quad = \cos(-3\pi/4)$

e) $\sin(11\pi/6) = \cos(\pi/2 - 11\pi/6)$

$\qquad\qquad\quad = \cos(-4\pi/3)$

Now, we can use Equation (2), letting $v = \pi/2 - u$, to obtain

$\sin(v) = \sin(\pi/2 - u)$

$\qquad\quad = \cos(\pi/2 - (\pi/2 - u))$

$\qquad\quad = \cos(u)\,.$

Therefore, we have

$$\sin(\pi/2 - u) = \cos(u) \qquad\qquad\qquad (3)$$

which is true for all real values of u.

EXERCISE 9.5

Rewrite each of the following in equivalent form using the sine function.

a) $\cos(-5\pi/3)$

b) $\cos(3\pi/4)$

c) $\cos(3\pi/2)$

d) $\cos(5\pi/6)$

e) $\cos(-7\pi/4)$

SOLUTION 9.5

a) $\cos(-5\pi/3) = \sin(\pi/2 - (-5\pi/3))$

$\qquad\qquad\qquad = \sin(\pi/2 + 5\pi/3)$

$\qquad\qquad\qquad = \sin(13\pi/6)$

b) $\cos(3\pi/4) = \sin(\pi/2 - 3\pi/4)$

$\qquad\qquad\quad = \sin(-\pi/4)$

c) $\cos(3\pi/2) = \sin(\pi/2 - 3\pi/2)$

$\qquad\qquad\quad = \sin(-\pi)$

d) $\cos(5\pi/6) = \sin(\pi/2 - 5\pi/6)$

$\qquad\qquad\quad = \sin(-\pi/3)$

e) $\cos(-7\pi/4) = \sin(\pi/2 - (-7\pi/4))$

$= \sin(\pi/2 + 7\pi/4)$

$= \sin(9\pi/4)$

Referring back to Equation (1), we can let $u = 0$ and obtain

$\cos(u - v) = \cos(u)\cos(v) + \sin(u)\sin(v)$

$\cos(0 - v) = \cos(0)\cos(v) + \sin(0)\sin(v)$

$= 1 \cdot \cos(v) + 0 \cdot \sin(v)$

$= \cos(v) + 0$

$= \cos(v) .$

Therefore, we have

$$\cos(-v) = \cos(v) \qquad (4)$$

which is true for all real values of v.

In a similar manner, we observe that

$\sin(-v) = \cos(\pi/2 - (-v))$

$= \cos(\pi/2 + v)$

$= \cos(v - (-\pi/2))$

$= \cos(v)\cos(-\pi/2) + \sin(v)\sin(-\pi/2)$

$= (\cos(v))(0) + (\sin(v))(-1)$

$= 0 - \sin(v)$

$= -\sin(v) .$

Therefore, we have

$$\sin(-v) = -\sin(v) \qquad (5)$$

which is true for all real values of v.

Using Equations (4) and (5), we conclude the following:

$$\begin{aligned}
\tan(-v) &= -\tan(v) \\
\cot(-v) &= -\cot(v) \\
\sec(-v) &= \sec(v) \\
\csc(-v) &= -\csc(v)
\end{aligned} \qquad (6)$$

which are true for all values of v for which the respective functions are defined.

We now determine that

$$\begin{aligned}
\cos(u+v) &= \cos(u-(-v)) \\
&= \cos(u)\cos(-v) + \sin(u)\sin(-v) \\
&= \cos(u)\cos(v) + \sin(u)(-\sin(v)),
\end{aligned}$$

from which we obtain

$$\cos(u+v) = \cos(u)\cos(v) - \sin(u)\sin(v) \qquad (7)$$

which is true for all real values of u and v.

EXERCISE 9.6

Determine the value of $\cos(5\pi/12)$.

SOLUTION 9.6

Since $5\pi/12 = \pi/4 + \pi/6$, we have

$$\begin{aligned}
\cos(5\pi/12) &= \cos(\pi/4 + \pi/6) \\
&= \cos(\pi/4)\cos(\pi/6) - \sin(\pi/4)\sin(\pi/6) \\
&= (\sqrt{2}/2)(\sqrt{3}/2) - (\sqrt{2}/2)(1/2) \\
&= \sqrt{6}/4 - \sqrt{2}/4 \\
&= \frac{\sqrt{6} - \sqrt{2}}{4}
\end{aligned}$$

EXERCISE 9.7

Determine the value of $\cos(5\pi/6)$.

SOLUTION 9.7

Since $5\pi/6 = \pi/2 + \pi/3$, we have

$$\cos(5\pi/6) = \cos(\pi/2 + \pi/3)$$

$$= \cos{(\pi/2)}\cos{(\pi/3)} - \sin{(\pi/2)}\sin{(\pi/3)}$$
$$= (0)(1/2) - (1)(\sqrt{3}/2)$$
$$= 0 - \sqrt{3}/2$$
$$= -\sqrt{3}/2$$

EXERCISE 9.8

Determine the value of $\cos{(13\pi/12)}$.

SOLUTION 9.8

Since $13\pi/12 = 3\pi/4 + \pi/3$, we have

$$\cos{(13\pi/12)} = \cos{(3\pi/4 + \pi/3)}$$
$$= \cos{(3\pi/4)}\cos{(\pi/3)} - \sin{(3\pi/4)}\sin{(\pi/3)}$$
$$= (-\sqrt{2}/2)(1/2) - (\sqrt{2}/2)(\sqrt{3}/2)$$
$$= -\sqrt{2}/4 - \sqrt{6}/2$$
$$= \frac{-(\sqrt{2} + \sqrt{6})}{4}$$

There are similar formulas for $\sin{(u+v)}$ and $\sin{(u-v)}$ that we now derive using the previously derived formulas. To obtain a formula for $\sin{(u+v)}$, we observe that

$$\sin{(u+v)} = \cos{(\pi/2 - (u+v))}$$
$$= \cos{((\pi/2 - u) - v)}$$
$$= \cos{(\pi/2 - u)}\cos{(v)} + \sin{(\pi/2 - u)}\sin{(v)}.$$

But, since

$$\cos{(\pi/2 - u)} = \sin{(u)}$$

and

$$\sin{(\pi/2 - u)} = \cos{(u)},$$

we have

$$\boxed{\sin{(u+v)} = \sin{(u)}\cos{(v)} + \cos{(u)}\sin{(v)}} \qquad (8)$$

which is true for all real values of u and v.

EXERCISE 9.9

Determine the value of $\sin{(7\pi/6)}$.

SOLUTION 9.9

Since $7\pi/6 = 2\pi/3 + \pi/2$, we have

$$
\begin{aligned}
\sin(7\pi/6) &= \sin(2\pi/3 + \pi/2) \\
&= \sin(2\pi/3)\cos(\pi/2) + \cos(2\pi/3)\sin(\pi/2) \\
&= (\sqrt{3}/2)(0) + (-1/2)(1) \\
&= 0 - 1/2 \\
&= -1/2
\end{aligned}
$$

EXERCISE 9.10

Determine the value of $\sin(13\pi/12)$.

SOLUTION 9.10

Since $13\pi/12 = 3\pi/4 + \pi/3$, we have

$$
\begin{aligned}
\sin(13\pi/12) &= \sin(3\pi/4 + \pi/3) \\
&= \sin(3\pi/4)\cos(\pi/3) + \cos(3\pi/4)\sin(\pi/3) \\
&= (\sqrt{2}/2)(1/2) + (-\sqrt{2}/2)(\sqrt{3}/2) \\
&= \sqrt{2}/4 - \sqrt{6}/4 \\
&= \frac{\sqrt{2} - \sqrt{6}}{4}
\end{aligned}
$$

EXERCISE 9.11

Determine the value of $\sin(11\pi/6)$.

SOLUTION 9.11

Since $11\pi/6 = 7\pi/6 + 2\pi/3$, we have

$$
\begin{aligned}
\sin(11\pi/6) &= \sin(7\pi/6 + 2\pi/3) \\
&= \sin(7\pi/6)\cos(2\pi/3) + \cos(7\pi/6)\sin(2\pi/3) \\
&= (-1/2)(-1/2) + (-\sqrt{3}/2)(\sqrt{3}/2) \\
&= 1/4 - 3/4 \\
&= -1/2
\end{aligned}
$$

Also,

$$
\begin{aligned}
\sin(u - v) &= \sin(u + (-v)) \\
&= \sin(u)\cos(-v) + \cos(u)\sin(-v) \\
&= \sin(u)\cos(v) + \cos(u)(-\sin(v)),
\end{aligned}
$$

from which we obtain

$$\sin (u - v) = \sin (u) \cos (v) - \cos (u) \sin (v) \qquad (9)$$

which is true for all real values of u and v.

EXERCISE 9.12

Determine the value of $\sin (5\pi/3)$.

SOLUTION 9.12

Since $5\pi/3 = 2\pi - \pi/3$, we have

$$
\begin{aligned}
\sin (5\pi/3) &= \sin (2\pi - \pi/3) \\
&= \sin (2\pi) \cos (\pi/3) - \cos (2\pi) \sin (\pi/3) \\
&= (0) (1/2) - (1) (\sqrt{3}/2) \\
&= 0 - \sqrt{3}/2 \\
&= -\sqrt{3}/2
\end{aligned}
$$

EXERCISE 9.13

Determine the value of $\sin (\pi/12)$.

SOLUTION 9.13

Since $\pi/12 = \pi/4 - \pi/6$, we have

$$
\begin{aligned}
\sin (\pi/12) &= \sin (\pi/4 - \pi/6) \\
&= \sin (\pi/4) \cos (\pi/6) - \cos (\pi/4) \sin (\pi/6) \\
&= (\sqrt{2}/2) (\sqrt{3}/2) - (\sqrt{2}/2) (1/2) \\
&= \sqrt{6}/4 - \sqrt{2}/4 \\
&= \frac{\sqrt{6} - \sqrt{2}}{4}
\end{aligned}
$$

EXERCISE 9.14

Determine the value of $\sin (5\pi/12)$.

SOLUTION 9.14

Since $5\pi/12 = 2\pi/3 - \pi/4$, we have

$$
\begin{aligned}
\sin (5\pi/12) &= \sin (2\pi/3 - \pi/4) \\
&= \sin (2\pi/3) \cos (\pi/4) - \cos (2\pi/3) \sin (\pi/4) \\
&= (\sqrt{3}/2) (\sqrt{2}/2) - (-1/2) (\sqrt{2}/2) \\
&= \sqrt{6}/4 + \sqrt{2}/4
\end{aligned}
$$

$$= \frac{\sqrt{6} + \sqrt{2}}{4}$$

To determine a formula for $\tan(u + v)$, we start with a fundamental identity

$$\tan(u + v) = \frac{\sin(u + v)}{\cos(u + v)}$$

and expand $\sin(u + v)$ and $\cos(u + v)$, obtaining

$$\tan(u + v) = \frac{\sin(u)\cos(v) + \cos(u)\sin(v)}{\cos(u)\cos(v) - \sin(u)\sin(v)}.$$

Next, we divide both the numerator and the denominator of this last equation by $\cos(u)\cos(v)$, provided that $\cos(u) \neq 0$ and $\cos(v) \neq 0$. Therefore,

$$\tan(u + v) = \frac{\dfrac{\sin(u)\cos(v)}{\cos(u)\cos(v)} + \dfrac{\cos(u)\sin(v)}{\cos(u)\cos(v)}}{\dfrac{\cos(u)\cos(v)}{\cos(u)\cos(v)} - \dfrac{\sin(u)\sin(v)}{\cos(u)\cos(v)}}$$

from which we obtain

$$\tan(u + v) = \frac{\tan(u) + \tan(v)}{1 - \tan(u)\tan(v)} \qquad (10)$$

which is true for all values of u and v for which the respective functions are defined.

Using a similar procedure, we obtain the following formula for $\tan(u - v)$.

$$\tan(u - v) = \frac{\tan(u) - \tan(v)}{1 + \tan(u)\tan(v)} \qquad (11)$$

EXERCISE 9.15

Determine the value for $\tan(5\pi/12)$.

SOLUTION 9.15

Since $5\pi/12 = \pi/4 + \pi/6$, we have

$$\tan(5\pi/12) = \tan(\pi/4 + \pi/6)$$

$$= \frac{\tan{(\pi/4)} + \tan{(\pi/6)}}{1 - \tan{(\pi/4)} \tan{(\pi/6)}}$$

$$= \frac{1 + \sqrt{3}/3}{1 - (1)(\sqrt{3}/3)}$$

$$= \frac{3 + \sqrt{3}}{3 - \sqrt{3}}$$

EXERCISE 9.16

Determine the value for $\tan{(\pi/12)}$.

SOLUTION 9.16

Since $\pi/12 = \pi/4 - \pi/6$, we have

$$\tan{(\pi/12)} = \tan{(\pi/4 - \pi/6)}$$

$$= \frac{\tan{(\pi/4)} - \tan{(\pi/6)}}{1 + \tan{(\pi/4)} \tan{(\pi/6)}}$$

$$= \frac{1 - \sqrt{3}/3}{1 + (1)(\sqrt{3}/3)}$$

$$= \frac{3 - \sqrt{3}}{3 + \sqrt{3}}$$

EXERCISE 9.17

Rewrite each of the following expressions as a single term.

a) $\cos{(4)} \cos{(7)} - \sin{(4)} \sin{(7)}$

b) $\sin{(3)} \cos{(5)} + \cos{(3)} \sin{(5)}$

c) $\sin{(-2)} \sin{(7)} - \cos{(2)} \cos{(7)}$

SOLUTION 9.17

a) $\cos{(4)} \cos{(7)} - \sin{(4)} \sin{(7)}$ is of the form $\cos{(u)} \cos{(v)} -$
   $\sin{(u)} \sin{(v)}$, which is equal to $\cos{(u + v)}$. Hence,

$$\cos{(4)} \cos{(7)} - \sin{(4)} \sin{(7)} = \cos{(4 + 7)}$$
$$= \cos{(11)}$$

b) $\sin{(3)} \cos{(5)} + \cos{(3)} \sin{(5)}$ is of the form $\sin{(u)} \cos{(v)} +$
   $\cos{(u)} \sin{(v)}$, which is equal to $\sin{(u + v)}$. Hence,

$$\sin{(3)} \cos{(5)} + \cos{(3)} \sin{(5)} = \sin{(3 + 5)}$$
$$= \sin{(8)}$$

c) $\sin(-2)\sin(7) - \cos(2)\cos(7)$

$$= -\sin(2)\sin(7) - \cos(2)\cos(7)$$

$$\text{(Since } \sin(-u) = -\sin(u))$$

$$= -[\sin(2)\sin(7) + \cos(2)\cos(7)]$$

$$= -[\cos(2)\cos(7) + \sin(2)\sin(7)]$$

$$= -\cos(2 - 7)$$

$$= -\cos(-5)$$

$$= -\cos(5) \qquad \text{(Since } \cos(-u) = \cos(u))$$

EXERCISE 9.18

Show that each of the following is a true statement for the permissible values of t.

a) $\cos(\pi - t) = -\cos(t)$

b) $\sin(t + \pi) = -\sin(t)$

c) $\tan(\pi - t) = -\tan(t)$

SOLUTION 9.18

a) $\cos(\pi - t) = \cos(\pi)\cos(t) + \sin(\pi)\sin(t)$

$$= (-1)\cos(t) + (0)\sin(t)$$

$$= -\cos(t) + 0$$

$$= -\cos(t)$$

b) $\sin(t + \pi) = \sin(t)\cos(\pi) + \cos(t)\sin(\pi)$

$$= \sin(t)(-1) + \cos(t)(0)$$

$$= -\sin(t) + 0$$

$$= -\sin(t)$$

c) $\tan(\pi - t) = \dfrac{\tan(\pi) - \tan(t)}{1 + \tan(\pi)\tan(t)}$

$$= \dfrac{0 - \tan(t)}{1 + (0)\tan(t)}$$

$$= \dfrac{-\tan(t)}{1 + 0}$$

$$= \dfrac{-\tan(t)}{1}$$

$$= -\tan(t)$$

EXERCISE 9.19

Determine the value for $\sin(u+v)$ if $W(u)$ lies in Quadrant I, with $\sin(u) = 1/3$, and $W(v)$ lies in Quadrant II, with $\sin(v) = 2/3$.

SOLUTION 9.19

If $\sin(u) = 1/3$, then $\cos^2(u) + \sin^2(u) = 1$ becomes

$$\cos^2(u) + (1/3)^2 = 1$$
$$\cos^2(u) + 1/9 = 1$$
$$\cos^2(u) = 8/9$$
$$\cos(u) = \pm\sqrt{8/9}$$
$$\cos(u) = \pm 2\sqrt{2}/3$$

But, since $W(u)$ lies in Quadrant I, $\cos(u) > 0$. Hence,

$$\cos(u) = 2\sqrt{2}/3.$$

If $\sin(v) = 2/3$, then $\cos^2(v) + \sin^2(v) = 1$ becomes

$$\cos^2(v) + (2/3)^2 = 1$$
$$\cos^2(v) + 4/9 = 1$$
$$\cos^2(v) = 5/9$$
$$\cos(v) = \pm\sqrt{5/9}$$
$$\cos(v) = \pm\sqrt{5}/3$$

But, since $W(v)$ lies in Quadrant II, $\cos(v) < 0$. Hence,

$$\cos(v) = -\sqrt{5}/3.$$

We now have

$$\sin(u+v) = \sin(u)\cos(v) + \cos(u)\sin(v)$$
$$= (1/3)(-\sqrt{5}/3) + (2\sqrt{2}/3)(2/3)$$
$$= -\sqrt{5}/9 + 4\sqrt{2}/9$$
$$= \frac{4\sqrt{2} - \sqrt{5}}{9}.$$

EXERCISE 9.20

Determine the value for $\tan(u-v)$ if $W(u)$ lies in Quadrant IV, with $\sin(u) = -8/17$, and $W(v)$ lies in Quadrant I, with $\tan(v) = 3/4$.

SOLUTION 9.20

Since $\sin(u) = -8/17$, $\cos^2(u) + \sin^2(u) = 1$ becomes

$$\cos^2(u) + (-8/17)^2 = 1$$
$$\cos^2(u) + 64/289 = 1$$
$$\cos^2(u) = 225/289$$
$$\cos(u) = \pm\sqrt{225/289}$$
$$\cos(u) = \pm 15/17$$

But, since $W(u)$ lies in Quadrant IV, $\cos(u) > 0$. Hence,

$$\cos(u) = 15/17.$$

We now have that

$$\tan(u) = \frac{\sin(u)}{\cos(u)} = \frac{-8/17}{15/17} = \frac{-8}{17}$$

Substituting $-8/17$ for $\tan(u)$ and $3/4$ for $\tan(v)$, we have

$$\tan(u - v) = \frac{\tan(u) - \tan(v)}{1 + \tan(u)\tan(v)}$$

$$= \frac{(-8/17) - (3/4)}{1 + (-8/17)(3/4)}$$

$$= \frac{-83/68}{1 - 24/68}$$

$$= \frac{-83/64}{44/68}$$

$$= \frac{-83}{44}.$$

9.2 CIRCULAR FUNCTION VALUES FOR MULTIPLES OF t

Formula for sin(2t)   If $t = \pi/4$, then $\sin(t) = \sin(\pi/4) = \sqrt{2}/2$. Also, $\sin(2t) = \sin(\pi/2) = 1$. Clearly, then, $\sin(2t) \neq 2 \cdot \sin(t)$ since $1 \neq 2(\sqrt{2}/2) = \sqrt{2}$. Since

$$\sin(u + v) = \sin(u)\cos(v) + \cos(u)\sin(v)$$

is true for all real values of u and v, we may let $u = t$ and $v = t$, and obtain

$$\sin(t+t) = \sin(t)\cos(t) + \cos(t)\sin(t).$$

This last equation simplifies to

$$\sin(2t) = 2 \cdot \sin(t)\cos(t) \tag{12}$$

which is true for all real values of t.

EXERCISE 9.21

Determine the value of $\sin(\pi/3)$ in terms of circular function values of $\pi/6$.

SOLUTION 9.21

Since $\pi/3 = 2(\pi/6)$, we use Equation (12) and obtain

$$\begin{aligned}
\sin(\pi/3) &= 2 \cdot \sin(\pi/6)\cos(\pi/6) \\
&= 2(1/2)(\sqrt{3}/2) \\
&= \sqrt{3}/2
\end{aligned}$$

which we already knew to be the correct value.

EXERCISE 9.22

Determine the value of $\sin(\pi/2)$ in terms of circular function values of $\pi/4$.

SOLUTION 9.22

Since $\pi/2 = 2(\pi/4)$, we use Equation (12) and obtain

$$\begin{aligned}
\sin(\pi/2) &= 2 \cdot \sin(\pi/4)\cos(\pi/4) \\
&= 2(\sqrt{2}/2)(\sqrt{2}/2) \\
&= 1
\end{aligned}$$

EXERCISE 9.23

Rewrite $\sin(\pi/8)$ in terms of circular function values of $\pi/16$.

SOLUTION 9.23

Since $\pi/8 = 2(\pi/16)$, we have

$$\sin(\pi/8) = 2 \cdot \sin(\pi/16)\cos(\pi/16)$$

Formulas for cos(2t)   In a similar manner, we may derive a formula for $\cos(2t)$. Starting with

$$\cos(u+v) = \cos(u)\cos(v) - \sin(u)\sin(v),$$

which is true for all real values of u and v, we may substitute $u = t$ and $v = t$ and obtain

$$\cos(t+t) = \cos(t)\cos(t) - \sin(t)\sin(t).$$

This last equation simplifies to

$$\cos(2t) = \cos^2(t) - \sin^2(t) \tag{13}$$

which is true for all real values of t.

Equation (13) may be written in two different but equivalent forms, using the identity $\cos^2(t) + \sin^2(t) = 1$. Since

$$\cos^2(t) = 1 - \sin^2(t),$$

we have

$$\cos(2t) = (1 - \sin^2(t)) - \sin^2(t)$$

or

$$\cos(2t) = 1 - 2 \cdot \sin^2(t) \tag{14}$$

which is true for all real values of t. Also, $\sin^2(t) = 1 - \cos^2(t)$. Hence, Equation (13) may take on the form

$$\cos(2t) = \cos^2(t) - (1 - \cos^2(t))$$

or

$$\cos(2t) = 2 \cdot \cos^2(t) - 1 \tag{15}$$

which is true for all real values of t.

EXERCISE 9.24

Determine the value for $\cos(2\pi/3)$ in terms of circular function values of $\pi/3$.

SOLUTION 9.24

Since $2\pi/3 = 2(\pi/3)$, we use Equation (13) and obtain

$$\cos{(2\pi/3)} = \cos^2{(\pi/3)} - \sin^2{(\pi/3)}$$
$$= (1/2)^2 - (\sqrt{3}/2)^2$$
$$= (1/4) - (3/4)$$
$$= -1/2$$

Or, we can use Equation (14) and obtain

$$\cos{(2\pi/3)} = 1 - 2 \cdot \sin^2{(\pi/3)}$$
$$= 1 - 2(\sqrt{3}/2)^2$$
$$= 1 - 2(3/4)$$
$$= 1 - 3/2$$
$$= -1/2$$

Or, we can use Equation (15) and obtain

$$\cos{(2\pi/3)} = 2 \cdot \cos^2{(\pi/3)} - 1$$
$$= 2(1/2)^2 - 1$$
$$= 2(1/4) - 1$$
$$= 1/2 - 1$$
$$= -1/2$$

EXERCISE 9.25

Evaluate $\cos{(\pi)}$ in terms of circular function values of $\pi/2$.

SOLUTION 9.25

Again, since $\pi = 2(\pi/2)$, we have a choice of using Equation (13), (14), or (15). We use Equation (13), obtaining

$$\cos{(\pi)} = \cos^2{(\pi/2)} - \sin^2{(\pi/2)}$$
$$= (0)^2 - (1)^2$$
$$= 0 - 1$$
$$= -1$$

EXERCISE 9.26

If $\sin{(t)} = -5/9$, determine a value for $\cos{(2t)}$.

SOLUTION 9.26

Using Equation (15), we have

$$\cos{(2t)} = 1 - 2 \cdot \sin^2{(t)}$$

$$= 1 - 2\,(-5/9)^2$$
$$= 1 - 2\,(25/81)$$
$$= 1 - 50/81$$
$$= 31/81$$

EXERCISE 9.27

Express $\cos{(4t)}$ in terms of circular function values of t.

SOLUTION 9.27

$$\cos{(4t)} = 2 \cdot \cos^2{(2t)} - 1$$
$$= 2\,(\cos{(2t)})^2 - 1$$
$$= 2\,(2 \cdot \cos^2{(t)} - 1)^2 - 1$$
$$= 2\,(4 \cdot \cos^4{(t)} - 4 \cdot \cos^2{(t)} + 1) - 1$$
$$= 8 \cdot \cos^4{(t)} - 8 \cdot \cos^2{(t)} + 2 - 1$$
$$= 8 \cdot \cos^4{(t)} - 8 \cdot \cos^2{(t)} + 1$$

Formula for tan(2t)   To obtain a formula for $\tan{(2t)}$ in terms of $\tan{(t)}$, we start with the equation

$$\tan{(u + v)} = \frac{\tan{(u)} + \tan{(v)}}{1 - \tan{(u)}\tan{(v)}}$$

which is true for all real values of u and v for which $\tan{(u)}$ and $\tan{(v)}$ are defined and is such that the denominator is not equal to zero. Now, if we let $u = t$ and $v = t$, the equation becomes

$$\tan{(t + t)} = \frac{\tan{(t)} + \tan{(t)}}{1 - \tan{(t)}\tan{(t)}}$$

which simplifies to

$$\tan{(2t)} = \frac{2 \cdot \tan{(t)}}{1 - \tan^2{(t)}} \tag{16}$$

which is true for all real values of t for which $\tan{(t)}$ is defined and such that $\tan{(t)} \neq \pm 1$.

EXERCISE 9.28

Evaluate $\tan{(\pi/3)}$ using circular function values of $\pi/6$.

SOLUTION 9.28

Since $\pi/3$ is equal to 2 times $\pi/6$, we let $t = \pi/6$ and $2t = \pi/3$. Hence, using Equation (16), we have

$$\tan(\pi/3) = \frac{2 \cdot \tan(\pi/6)}{1 - \tan^2(\pi/6)}$$

$$= \frac{2(\sqrt{3}/3)}{1 - (\sqrt{3}/3)^2}$$

$$= \frac{2\sqrt{3}/3}{1 - 1/3}$$

$$= \frac{2\sqrt{3}/3}{2/3}$$

$$= \sqrt{3}$$

EXERCISE 9.29

Evaluate $\tan(14\pi/9)$ in terms of circular function values of $7\pi/9$.

SOLUTION 9.29

$14\pi/9 = 2(7\pi/9)$. We use Equation (16), obtaining

$$\tan(14\pi/9) = \frac{2 \cdot \tan(7\pi/9)}{1 - \tan^2(7\pi/9)}$$

$sin^2(t)$ and $cos^2(t)$

Equations (14) and (15) may be rewritten as

$$\sin^2(t) = \frac{1 - \cos(2t)}{2} \qquad (17)$$

and

$$\cos^2(t) = \frac{1 + \cos(2t)}{2} \qquad (18)$$

Dividing both sides of Equation (17) by the corresponding sides of Equation (18), we obtain

$$\tan^2 (t) = \frac{1 - \cos (2t)}{1 + \cos (2t)} \qquad\qquad (19)$$

which is true for all real values of t for which $\cos (2t) \neq -1$.

EXERCISE 9.30

Evaluate $\sin (\pi/8)$.

SOLUTION 9.30

Since $\pi/8$ is one-half of $\pi/4$, we may let $t = \pi/8$ and $2t = \pi/4$. Hence, using Equation (17), we have

$$\sin^2 (\pi/8) = \frac{1 - \cos (\pi/4)}{2}$$

$$= \frac{1 - \sqrt{2}/2}{2}$$

$$= \frac{2 - \sqrt{2}}{4}$$

Since $W(\pi/8)$ lies in Quadrant I, $\sin (\pi/8) > 0$. Therefore, extracting the square root, we find that

$$\sin (\pi/8) = \frac{\sqrt{2 - \sqrt{2}}}{2},$$

which is the exact value for $\sin (\pi/8)$. Approximate values may be obtained by evaluating the square roots involved to a specified number of significant digits.

EXERCISE 9.31

Evaluate $\cos (11\pi/12)$.

SOLUTION 9.31

Since $11\pi/12$ is one-half of $11\pi/6$, we may let $t = 11\pi/12$ and $2t = 11\pi/6$. Hence, using Equation (18), we have

$$\cos^2 (11\pi/12) = \frac{1 + \cos (11\pi/6)}{2}$$

$$= \frac{1 + \sqrt{3}/2}{2}$$

$$= \frac{2 + \sqrt{3}}{4}$$

Since $W(11\pi/12)$ lies in Quadrant II, $\cos(11\pi/12) < 0$. Therefore, extracting the negative square root, we have

$$\cos(11\pi/12) = \frac{-\sqrt{2 + \sqrt{3}}}{2}$$

which is the exact value for $\cos(11\pi/12)$.

EXERCISE 9.32

Determine the exact value of $\tan(\pi/8)$.

SOLUTION 9.32

Since $\pi/8 = 1/2$ times $\pi/4$, we use Equation (19), obtaining

$$\tan^2(\pi/8) = \frac{1 - \cos(\pi/4)}{1 + \cos(\pi/4)}$$

$$= \frac{1 - \sqrt{2}/2}{1 + \sqrt{2}/2}$$

$$= \frac{2 - \sqrt{2}}{2 + \sqrt{2}}$$

$$= \frac{2 - \sqrt{2}}{2 + \sqrt{2}} \cdot \frac{2 - \sqrt{2}}{2 - \sqrt{2}}$$

$$= \frac{4 - 2\sqrt{2} + 2}{4 - 2}$$

$$= \frac{6 - 2\sqrt{2}}{2}$$

$$= 3 - 2\sqrt{2}$$

Since $W(\pi/8)$ lies in Quadrant I, $\tan(\pi/8) > 0$. Hence,

$$\tan(\pi/8) = \sqrt{3 - 2\sqrt{2}}$$

EXERCISE 9.33

Given $\sin(t) = 5/13$, with $W(t)$ in Quadrant I, determine values for:

a) $\sin(2t)$

b) $\sin(t/2)$

SOLUTION 9.33

Since $\sin(t) = 5/13$ and $W(t)$ lies in Quadrant I, we determine the value of $\cos(t)$ as follows:

$$\cos^2(t) + \sin^2(t) = 1$$
$$\cos^2(t) + (5/13)^2 = 1$$
$$\cos^2(t) + 25/169 = 1$$
$$\cos^2(t) = 144/169$$
$$\cos(t) = 12/13 \qquad \text{(Since } \cos(t) > 0\text{)}$$

a) Using Equation (12), we have

$$\sin(2t) = 2 \cdot \sin(t)\cos(t)$$
$$= 2(5/13)(12/13)$$
$$= 120/169$$

b) Using Equation (17), with t and $2t$ replaced by $t/2$ and t, respectively, we have

$$\sin^2(t/2) = \frac{1 - \cos(t)}{2}$$
$$= \frac{1 - 12/13}{2}$$
$$= \frac{1/13}{2}$$
$$= 1/26 \approx 0.038$$

Since $W(t)$ lies in Quadrant I, $W(t/2)$ will also lie in Quadrant I. Hence, $\sin(t/2) > 0$, and we have

$$\sin(t/2) = \sqrt{1/26} \approx 0.196$$

EXERCISE 9.34

Given $\cos(t) = -12/13$, with $W(t)$ in Quadrant II, determine values for

a) $\cos(2t)$

b) $\cos(t/2)$

SOLUTION 9.34

a) Using Equation (15), we have

$$\cos(2t) = 2 \cdot \cos^2(t) - 1$$
$$= 2(-12/13)^2 - 1$$

$$= 2(144/169) - 1$$

$$= 288/169 - 1$$

$$= 119/169 \approx 0.704$$

b) Using Equation (18), with t and $2t$ replaced by $t/2$ and t, respectively, we have

$$\cos^2(t/2) = \frac{1 + \cos(t)}{2}$$

$$= \frac{1 + (-12/13)}{2}$$

$$= \frac{1/13}{2}$$

$$= 1/26$$

Since $W(t)$ lies in Quadrant II, then $\pi/2 \le t \le \pi$. Therefore, $\pi/4 \le t/2 \le \pi/2$, and $W(t/2)$ lies in Quadrant I. Therefore, $\cos(t/2) > 0$, and $\cos(t/2) = \sqrt{1/26} \approx 0.196$.

EXERCISE 9.35

Given $\tan(t) = 0.4$, with $W(t)$ in Quadrant III, determine values for:

a) $\tan(2t)$

b) $\tan(t/2)$

SOLUTION 9.35

a) Using Equation (16), we have

$$\tan(2t) = \frac{2 \cdot \tan(t)}{1 - \tan^2(t)}$$

$$= \frac{2(0.4)}{1 - (0.4)^2}$$

$$= \frac{0.8}{1 - 0.16}$$

$$= \frac{0.8}{0.84}$$

$$= 2/21 \approx 0.952$$

b) Using Equation (19), with t and $2t$ replaced by $t/2$ and t, respectively, we have

$$\tan^2(t/2) = \frac{1 - \cos(t)}{1 + \cos(t)}$$

To determine the value of $\cos(t)$, we use the basic identity

$$1 + \tan^2(t) = \sec^2(t)$$
$$1 + (0.4)^2 = \sec^2(t)$$
$$1 + 0.16 = \sec^2(t)$$
$$1.16 = \sec^2(t)$$

Since $W(t)$ is in Quadrant III, $\sec(t) < 0$. Hence,

$$\sec(t) = -\sqrt{1.16}$$

and

$$\cos(t) = \frac{-1}{\sqrt{1.16}}$$

Continuing, we have

$$\tan^2(t) = \frac{1 - \cos(t)}{1 + \cos(t)}$$

$$= \frac{1 - (-1/\sqrt{1.16})}{1 + (-1/\sqrt{1.16})}$$

$$= \frac{1 + 1/\sqrt{1.16}}{1 - 1/\sqrt{1.16}}$$

$$\approx 26.963$$

$$\tan\left(\frac{t}{2}\right) \approx \pm\sqrt{26.963} \approx \pm 5.19$$

Since $w(t)$ is in Quadrant III, $w(t/2)$ is in Quadrant II and $\tan(t/2) < 0$. Hence $\tan(t/2) \approx -5.19$.

9.3 *Graphs of $y = a \cdot sin(bx + c)$ and $y = a \cdot cos(bx + c)$*

The circular functions that we have examined so far are very interesting in their basic form. However, from the viewpoint of the engineer, the physicist, the technologist, and the businessman, in certain areas, variations of these functions become even more interesting. We will now consider such variations by examining the graphs associated with the equations $y = a \cdot \sin(bx + c)$ and $y = a \cdot \cos(bx + c)$ where a, b, and c are constants with $a \neq 0$.

Stretching and Shrinking

Let us first look at $y = a \cdot \sin(x)$ and $y = a \cdot \cos(x)$. From chapter 4, recall that the function g, defined by $g(x) = cf(x)$ for all x in the domain of f, is determined as follows:

- If $|c| > 1$, the graph of g is a vertical stretching of the graph of f away from the x-axis.
- If $0 < |c| < 1$, the graph of g is a vertical shrinking of the graph of f toward the x-axis.
- If $c = -1$, the graph of g is a reflection of the graph of f about the x-axis.

Hence, the graphs of $y = a \cdot \sin(x)$ and $y = a \cdot \cos(x)$ can be obtained from the graphs of $y = \sin(x)$ and $y = \cos(x)$ simply by multiplying all of the ordinates of the points on the basic graphs by a. Note, however, that if $a < 0$, we would also have a reflection about the x-axis.

EXERCISE 9.36

Graph the function defined by $y = 2 \cdot \sin(x)$.

SOLUTION 9.36

First, we graph the sin function on the interval $[0, 2\pi]$. Then, the graph of $y = 2 \cdot \sin(x)$ would be stretched vertically by a factor of 2. (See Figure 9.2.) Observe that the real zeros of $\sin(x)$ and $2 \cdot \sin(x)$ are the same. Since the graph of $y = \sin(x)$ is periodic with a period of 2π, the graph of $y = 2 \cdot \sin(x)$ is also periodic with a period of 2π.

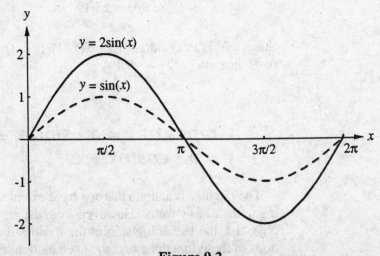

Figure 9.2
The Graphs of $y = \sin(x)$ and $y = 2 \cdot \sin(x)$ on the Interval $[0, 2\pi]$

EXERCISE 9.37

Graph the function defined by $y = 3 \cdot \cos(x)$.

SOLUTION 9.37

First, we graph the cos function on the interval $[0, 2\pi]$. Then, the graph of $y = 3 \cdot \cos(x)$ would be stretched vertically by a factor of 3. (See Figure 9.3.) Observe that the real zeros of $\cos(x)$ and $3 \cdot \cos(x)$ are the same. Again, since the graph of $y = \cos(x)$ is periodic with a period of 2π, the graph of $y = 3 \cdot \cos(x)$ is also periodic with a period of 2π.

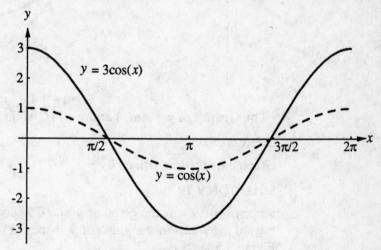

Figure 9.3

The Graphs of $y = \cos(x)$ and $y = 3 \cdot \cos(x)$ on the Interval $[0, 2\pi]$

EXERCISE 9.38

Graph the function defined by $y = (1/2) \sin(x)$.

SOLUTION 9.38

We proceed as in the previous exercises, noting that $a = 1/2 < 1$. Hence, the graph of $y = (1/2) \sin(x)$ is a vertical shrinking toward the x-axis of the graph of $\sin(x)$, with a factor of $1/2$. (See Figure 9.4.)

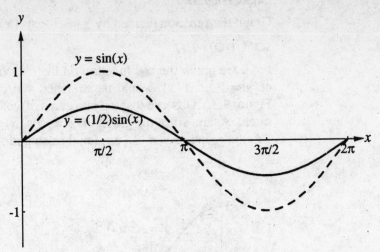

Figure 9.4
The Graphs of $y = \sin(x)$ and $y = (1/2)\sin(x)$ on the Interval $[0, 2\pi]$

EXERCISE 9.39

Graph the function defined by $y = (0.7) \cos (x)$.

SOLUTION 9.39

Since $a = 0.7 < 1$, the graph of $y = (0.7) \cos (x)$ is a vertical shrinking toward the x-axis of the graph of $y = \cos (x)$, with a factor of 0.7. (See Figure 9.5.)

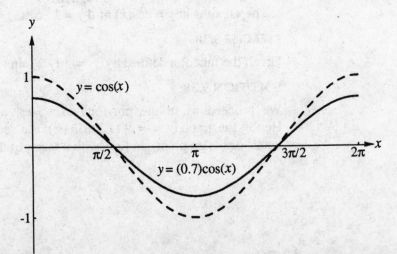

Figure 9.5
The Graphs of $y = \cos(x)$ and $y = (0.7)\cos(x)$ on the Interval $[0, 2\pi]$

EXERCISE 9.40

Graph the function defined by $y = -3\sin(x)$.

SOLUTION 9.40

The solution is obtained in two steps. First, start with the graph of $y = \sin(x)$. Then, stretch it vertically with a factor of 3, to obtain the graph of $y = 3 \cdot \sin(x)$. (See Figure 9.6.) Then, reflect the graph of $y = 3 \cdot \sin(x)$ about the x-axis to obtain the graph of $y = -3 \cdot \sin(x)$. (See Figure 9.7.)

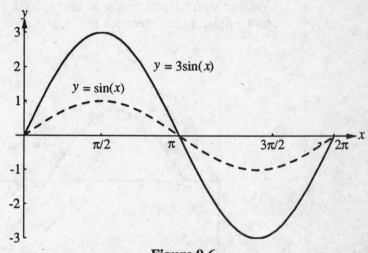

Figure 9.6

The Graphs of $y = \sin(x)$ and $y = 3 \cdot \sin(x)$ on the Interval $[0, 2\pi]$

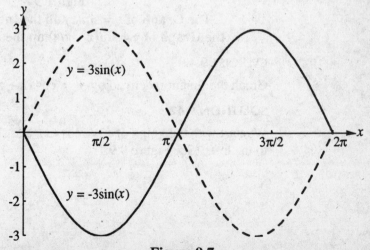

Figure 9.7

The Graphs of $y = 3 \cdot \sin(x)$ and $y = -3 \cdot \sin(x)$ on the Interval $[0, 2\pi]$

Horizontal Shifts

Next, we consider the graphs of $y = \sin(x + c)$ and $y = \cos(x + c)$. Again, recall from chapter 4 that the graph of g defined by $g(x) = f(x - c)$ is a horizontal shift of the graph of f, with a shift of $|c|$ units. If $c > 0$, the shift is to the right. If $c < 0$, the shift is to the left.

EXERCISE 9.41

Graph the function defined by $y = \sin(x - \pi/4)$.

SOLUTION 9.41

We start with the graph of $y = \sin(x)$ and shift it horizontally $\pi/4$ units to the right. (See Figure 9.8.)

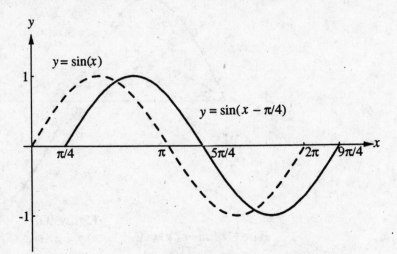

Figure 9.8
The Graph of $y = \sin(x)$ on the Interval $[0, 2\pi]$ and
the Graph of $y = \sin(x - \pi/4)$ on the Interval $[\pi/4, 9\pi/4]$

EXERCISE 9.42

Graph the function defined by $y = \cos(x + \pi/3)$.

SOLUTION 9.42

We start with the graph of $y = \cos(x)$ and shift it horizontally $\pi/3$ units to the left. (See Figure 9.9.)

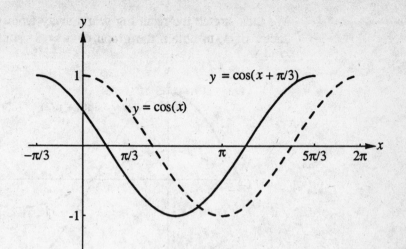

Figure 9.9
The Graph of $y = \cos(x)$ on the Interval $[0, 2\pi]$ and
the Graph of $y = \cos(x + \pi/3)$ on the Interval $[-\pi/3, 5\pi/3]$

EXERCISE 9.43

Graph the function defined by $y = 3 \cdot \sin(x + \pi/2)$.

SOLUTION 9.43

We start with the graph of $y = \sin(x)$ and shift it horizontally $\pi/2$ units to the left, obtaining the graph of $y = \sin(x + \pi/2)$. (See Figure 9.10.)

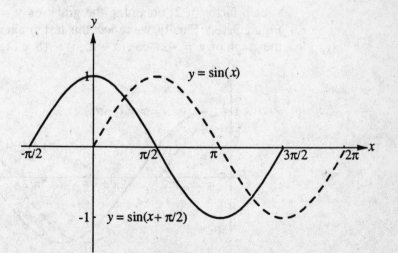

Figure 9.10
The Graph of $y = \sin(x)$ on the Interval $[0, 2\pi]$ and
the Graph of $y = \sin(x + \pi/2)$ on the Interval $[-\pi/2, 3\pi/2]$

We then stretch the resulting graph away from the *x*-axis, with a stretch factor of 3, to obtain the graph of $y = 3 \cdot \sin(x + \pi/2)$. (See Figure 9.11.)

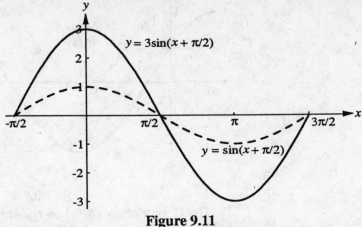

Figure 9.11
The Graphs of $y = \sin(x + \pi/2)$ and $y = 3 \cdot \sin(x + \pi/2)$ on the Interval $[-\pi/2, 3\pi/2]$

EXERCISE 9.44

Graph the function defined by $y = -2 \cdot \cos(x - \pi/6)$.

SOLUTION 9.44

We start with the graph of $y = \cos(x)$ and shift it horizontally $\pi/6$ units to the right, obtaining the graph of $y = \cos(x - \pi/6)$. (See Figure 9.12.) Next, we stretch the resulting graph away from the *x*-axis, with a stretch factor of 2, obtaining the graph of $y = 2 \cdot \cos(x - \pi/6)$. (See Figure 9.13.) Finally, we reflect this last graph about the *x*-axis, obtaining the graph of $y = -2 \cdot \cos(x - \pi/6)$. (See Figure 9.14.)

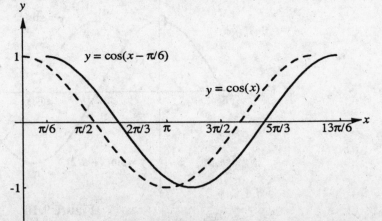

Figure 9.12
The Graph of $y = \cos(x)$ on the Interval $[0, 2\pi]$ and the Graph of $y = \cos(x - \pi/6)$ on the Interval $[\pi/6, 13\pi/6]$

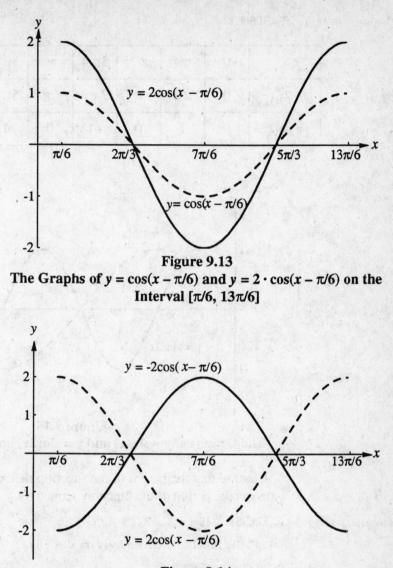

Figure 9.13
The Graphs of $y = \cos(x - \pi/6)$ and $y = 2 \cdot \cos(x - \pi/6)$ on the
Interval $[\pi/6, 13\pi/6]$

Figure 9.14
The Graphs of $y = 2 \cdot \cos(x - \pi/6)$ and $y = -2 \cdot \cos(x - \pi/6)$ on the
Interval $[\pi/6, 13\pi/6]$

sin(bx) and cos(bx)   We now consider the graphs of $y = \sin(bx)$ and $y = \cos(bx)$, with $b > 0$. For instance, to determine the graph of $y = \sin(2x)$, we plot the points indicated in Table 9.1. The graph is given in Figure 9.15.

TABLE 9.1

x	0	$\pi/4$	$\pi/2$	$3\pi/4$	π	$5\pi/4$	$3\pi/2$	$7\pi/4$	2π
$2x$	0	$\pi/2$	π	$3\pi/2$	2π	$5\pi/2$	3π	$7\pi/2$	4π
$y = \sin(2x)$	0	1	0	−1	0	1	0	−1	0

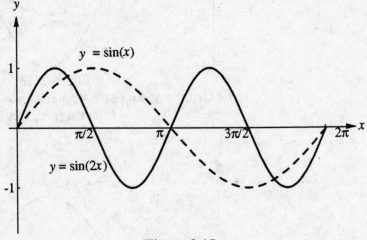

Figure 9.15
The Graphs of $y = \sin(x)$ and $y = \sin(2x)$ on the Interval $[0, 2\pi]$

Observe that the period of the function defined by $y = \sin(2x)$ is π, whereas the period of the function defined by $y = \sin(x)$ is 2π.

EXERCISE 9.45

Graph the function defined by $y = \cos(3x)$.

SOLUTION 9.45

To graph $y = \cos(3x)$, we plot the points indicated in Table 9.2. The graph is given in Figure 9.16.

TABLE 9.2

x	0	$\pi/6$	$\pi/3$	$\pi/2$	$2\pi/3$	$5\pi/6$	π	$7\pi/6$	$4\pi/3$	$3\pi/2$	$5\pi/3$	$11\pi/6$	2π
$3x$	0	$\pi/2$	π	$3\pi/2$	2π	$5\pi/2$	3π	$7\pi/2$	4π	$9\pi/2$	5π	$11\pi/2$	6π
$y = \cos(3x)$	1	0	−1	0	1	0	−1	0	1	0	−1	0	1

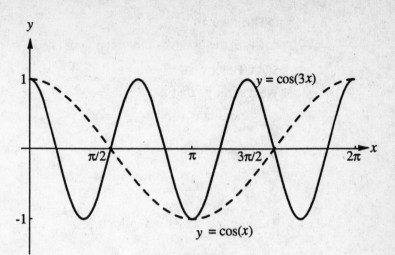

Figure 9.16
The Graphs of $y = \cos(x)$ and $y = \cos(3x)$ on the Interval $[0, 2\pi]$

Observe that the period of the function $y = \cos(3x)$ is $2\pi/3$, whereas the period of the function $y = \cos(x)$ is 2π.

> In general, if $b > 0$, then the period of the graphs of $y = \sin(bx)$ and $y = \cos(bx)$ is equal to $2\pi/b$.

Amplitude, Shifts, and Period

Let f and g be two functions defined by $y = f(x) = a \cdot \sin(bx + c)$ and $y = g(x) = a \cdot \cos(bx + c)$.

Amplitude: The amplitude of the graphs of f and g is given by $|a|$ and indicates the factor of stretch or shrink of the graphs of sin and cos required to obtain the graphs of f and g, respectively.

Phase Shift: If $b = 1$, the constant c is called the phase shift for the graphs of f and g and indicates the amount of horizontal shift of the graphs of sin and cos required to obtain the graphs of f and g, respectively.

a) If $c > 0$, the phase shift is to the left.

b) If $c < 0$, the phase shift is to the right.

If $b \neq 1$, we factor $bx + c$ to obtain $b(x + c/b)$. The phase shift, then, is equal to $\dfrac{|c|}{b}$ units.

Period: The constant b determines the period of the graphs of f and g. The period is equal to $\dfrac{2\pi}{|b|}$.

EXERCISE 9.46

Analyze the graph of the function defined by $y = 2 \cdot \cos(3x - \pi/6)$.

SOLUTION 9.46

We determine that $a = 2$, $b = 3$, and $c = -\pi/6$.

Amplitude $= |a| = |2| = 2$.

Phase shift is $\dfrac{|c|}{b} = \dfrac{|-\pi/6|}{3} = \dfrac{\pi}{18}$ units to the right.

Period is $\dfrac{2\pi}{b} = \dfrac{2\pi}{3}$.

The graph is found in Figure 9.17.

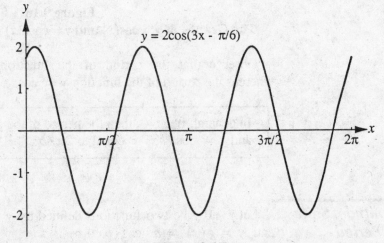

Figure 9.17

The Graph of $y = 2 \cdot \cos(3x - \pi/6)$ on the Interval $[0, 2\pi]$

EXERCISE 9.47

Analyze the graph of the function defined by $y = 3 \cdot \sin\left(\dfrac{1}{2}x - \dfrac{\pi}{4}\right)$.

SOLUTION 9.47

We determine that $a = 3$, $b = 1/2$, and $c = -\pi/4$.

Amplitude $= |a| = |3| = 3$.

Phase shift is $\dfrac{|c|}{b} = \dfrac{|-\pi/4|}{1/2} = \dfrac{\pi}{2}$ units to the right.

Period is $\dfrac{2\pi}{b} = \dfrac{2\pi}{1/2} = 4\pi$.

The graph of the function is found in Figure 9.18.

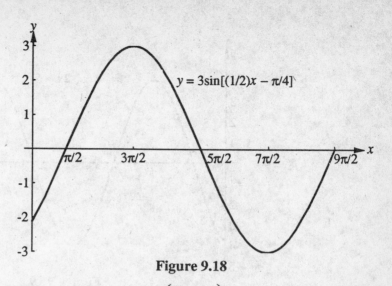

Figure 9.18

The Graph of $y = 3 \cdot \sin\left(\dfrac{1}{2}x - \dfrac{\pi}{4}\right)$ on the Interval $[0, 9\pi/4]$

EXERCISE 9.48

Analyze the graph of the function defined by $y = -4 \cdot \cos(-2x + 6)$.

SOLUTION 9.48

Rewriting, we have

$$y = -4 \cdot \cos(-2x + 6)$$
$$= -4 \cdot \cos[-(2x - 6)]$$
$$= -4 \cdot \cos(2x - 6) \qquad\qquad (\text{Since } \cos(-t) = \cos(t))$$

with $a = -4$, $b = 2$, and $c = -6$.

Amplitude $= |a| = |-4| = 4$.

Phase shift is $\dfrac{|c|}{b} = \dfrac{|-6|}{2} = \dfrac{6}{2} = 3$ units to the right.

Period is $\dfrac{2\pi}{b} = \dfrac{2\pi}{2} = \pi$.

Also note that, since $a < 0$, there is a reflection about the x-axis to get the graph of the required function. The graph is given in Figure 9.19.

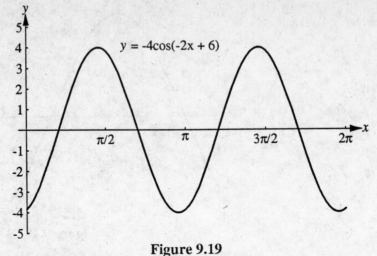

Figure 9.19
The Graph of $y = -4 \cdot \cos(-2x + 6)$ on the Interval $[0, 2\pi]$

EXERCISE 9.49

Analyze the graph of the function defined by $y = -3 \cdot \sin(-3x - 9)$.

SOLUTION 9.49

Rewriting, we have

$$y = -3 \cdot \sin(-3x - 9)$$
$$= -3 \cdot \sin[-(3x + 9)]$$
$$= 3 \cdot \sin(3x + 9) \qquad \text{(Since } \sin(-t) = -\sin(t)\text{)}$$

with $a = 3$, $b = 3$, and $c = 9$.

Amplitude = $|a| = |3| = 3$.

Phase shift is $\dfrac{|c|}{b} = \dfrac{|9|}{3} = 3$ units to the left.

Period is $\dfrac{2\pi}{b} = \dfrac{2\pi}{3}$.

The graph is given in Figure 9.20.

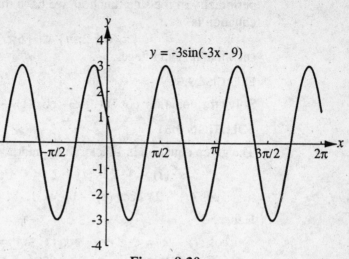

Figure 9.20
The Graph of $y = -3 \cdot \sin(-3x - 9)$ on the Interval $[-3, 2\pi]$

9.4 TRIGONOMETRIC EQUATIONS

A trigonometric equation is an equation involving one or more trigonometric or inverse trigonometric expressions. The solution set of a trigonometric equation may be the empty set (i.e., the set containing no elements), a set containing a finite number of elements, or a set containing infinitely many elements. For instance, since the maximum value that $\sin(t)$ may take is 1, the equation $\sin(t) = 2$ has no solutions. Hence, the solution set for this equation is the empty set, denoted by $\emptyset$. However, the equation $\cos(t) = 1/2$ such that $0 \le t \le \pi/2$ has a single solution, which is $\pi/3$ since $\cos(\pi/3) = 1/2$ and $0 \le \pi/3 \le \pi/2$. The solution set is $\{\pi/3\}$. Finally, the equation $\tan(t) = 1$ has infinitely many solutions, among which are $\pi/4$, $5\pi/4$, $-3\pi/4$, $-7\pi/4$, $9\pi/4$, $13\pi/4$, $-11\pi/4$, and $-15\pi/4$. Hence, the solution set is
$$\{\pi/4 \pm 2\pi n\} \cup \{5\pi/4 \pm 2\pi n\}$$
such that n is an integer.

EXERCISE 9.50

Solve the equation $\sin(t) = 1/2$ on the set $(-\infty, +\infty)$.

SOLUTION 9.50

Since $\sin(t)$ is positive, $W(t)$ lies in Quadrant I or Quadrant II. If $W(t)$ lies in Quadrant I, $t = \pi/6$ since $\sin(\pi/6) = 1/2$. If $W(t)$ lies in Quadrant II, then $t = 5\pi/6$ since $\sin(5\pi/6) = 1/2$. Using the periodicity of the sine function, we have that the solution set for the given equation is

$$\{\pi/6 \pm 2\pi n\} \cup \{5\pi/6 \pm 2\pi n\}$$

such that n is an integer.

EXERCISE 9.51

Solve the equation $\cos^2(t) - 3 \cdot \cos(t) + 2 = 0$ such that $0 \le t \le 2\pi$.

SOLUTION 9.51

The given equation is a factorable quadratic equation in $\cos(t)$. Hence,

$$\cos^2(t) - 3 \cdot \cos(t) + 2 = 0$$
$$(\cos(t) - 2)(\cos(t) - 1) = 0$$

Hence,

$$\cos(t) - 2 = 0 \quad \text{or} \quad \cos(t) - 1 = 0$$
$$\cos(t) = 2 \qquad\qquad \cos(t) = 1$$

But, the equation $\cos(t) = 2$ has no solution. The solutions of the equation $\cos(t) = 1$, on the given interval, are 0 and 2π. Hence, the solution set for the given equation, on the given interval, is $\{0, 2\pi\}$.

EXERCISE 9.52

Solve the equation $1 - 2 \cdot \sin^2(t) + \cos(t) = 0$ such that $0 \le t \le 4\pi$.

SOLUTION 9.52

The given equation involves two trigonometric functions. However, we may substitute $1 - \cos^2(t)$ for $\sin^2(t)$ and obtain

$$1 - 2(1 - \cos^2(t)) + \cos(t) = 0.$$

Simplifying, we have

$$2 \cdot \cos^2(t) + \cos(t) - 1 = 0$$
$$(2 \cdot \cos(t) - 1)(\cos(t) + 1) = 0$$

Therefore, either

$$2 \cdot \cos(t) - 1 = 0 \quad \text{or} \quad \cos(t) + 1 = 0$$
$$2 \cdot \cos(t) = 1 \qquad\qquad \cos(t) = -1$$
$$\cos(t) = 1/2$$

If $\cos(t) = 1/2$, then, on the given interval, t equals $\pi/3$, $5\pi/3$, $7\pi/3$, and $11\pi/3$. If $\cos(t) = -1$, then, on the given interval, t equals π or 3π. Hence, the required solution set is

$$\{\pi/3, \pi, 5\pi/3, 7\pi/3, 3\pi, 11\pi/3\}.$$

EXERCISE 9.53

Solve the equation $\sec^2(t) + \tan(t) = 0$ such that $0 \le t \le 3\pi$.

SOLUTION 9.53

Since $\sec^2(t) = 1 + \tan^2(t)$, the given equation may be rewritten as

$$1 + \tan^2(t) + \tan(t) = 0,$$

which is quadratic in $\tan(t)$. Using the Quadratic Formula, we have

$$\tan(t) = \frac{-1 \pm \sqrt{(1)^2 - 4(1)(1)}}{2} = \frac{-1 \pm \sqrt{1-4}}{2} = \frac{-1 \pm i\sqrt{3}}{2}$$

There are no real solutions for the given equation. Hence, the solution set for the given equation is the empty set, $\emptyset$.

EXERCISE 9.54

Solve the equation $\sec^2(t) - \sec^2(t)\sin^2(t) + 2 \cdot \cos^2(t) = 2$ such that $-\pi \le t \le 3\pi$.

SOLUTION 9.54

This equation involves three trigonometric functions. The approach taken here is to transform the equation into an equivalent equation involving a single trigonometric function, if possible. Using the identity $\sin^2(t) + \cos^2(t) = 1$, we substitute $1 - \cos^2(t)$ for $\sin^2(t)$, obtaining

$$\sec^2(t) - \sec^2(t)(1 - \cos^2(t)) + 2 \cdot \cos^2(t) = 2.$$

Expanding, we have

$$\sec^2(t) - \sec^2(t) + \sec^2(t)\cos^2(t) + 2 \cdot \cos^2(t) = 2$$
$$\sec^2(t)\cos^2(t) + 2 \cdot \cos^2(t) = 2$$

But, $\sec^2(t)\cos^2(t) = 1$. Hence,

$$\sec^2(t)\cos^2(t) + 2\cos^2(t) = 2$$
$$1 + 2\cos^2(t) = 2$$
$$2\cos^2(t) = 1$$
$$\cos^2(t) = 1/2$$
$$\cos(t) = \pm\sqrt{2}/2.$$

The solution set for the given equation, on the given interval, is
$$\{-3\pi/4, -\pi/4, \pi/4, 3\pi/4\}.$$

EXERCISE 9.55

Solve the equation $\cos(2t) = 3 \cdot \sin(t) + 2$ such that $0 \le t \le 2\pi$.

SOLUTION 9.55

The given equation involves t and $2t$. Using the formula

$$\cos(2t) = 1 - 2 \cdot \sin^2(t),$$

we rewrite the given equation in the equivalent form,

$$1 - 2\sin^2(t) = 3 \cdot \sin(t) + 2$$

or

$$2\sin^2(t) + 3 \cdot \sin(t) + 1 = 0,$$

which is quadratic in $\sin(t)$. We have

$$2\sin^2(t) + 3 \cdot \sin(t) + 1 = 0$$

$$(2 \cdot \sin(t) + 1)(\sin(t) + 1) = 0$$

from which we determine that either

$$2 \cdot \sin(t) + 1 = 0 \qquad \text{or} \qquad \sin(t) + 1 = 0$$

$$2 \cdot \sin(t) = -1 \qquad\qquad \sin(t) = -1$$

$$\sin(t) = -1/2$$

If $\sin(t) = -1/2$, then, on the given interval, t equals $7\pi/6$ or $11\pi/6$. If $\sin(t) = -1$, then, on the given interval, $t = 3\pi/2$. Hence, the solution set for the given equation, on the given interval, is $\{7\pi/6,\ 3\pi/2,\ 11\pi/6\}$.

EXERCISE 9.56

Solve the equation $\tan(t) - 1 = \sec(t)$ such that $0 \le t \le 2\pi$.

SOLUTION 9.56

The given equation involves two trigonometric functions, but using the fundamental identity,

$$1 + \tan^2(t) = \sec^2(t),$$

we may substitute

$$\pm\sqrt{1 + \tan^2(t)}$$

for $\sec(t)$ in the original equation, obtaining

$$\tan(t) - 1 = \pm\sqrt{1 + \tan^2(t)}.$$

Squaring both sides of this equation, we obtain

$$\tan^2(t) - 2 \cdot \tan(t) + 1 = 1 + \tan^2(t)$$

which is equivalent to

$$-2 \cdot \tan(t) = 0$$

or

$$\tan (t) = 0,$$

which is true, on the given interval, for t equal to 0, π, or 2π. All three of these values are solutions of the last equation. However, since we squared both sides of an equation, it is possible that we may have introduced an extraneous root. Hence, checking in the *original* equation, we determine:

If $t = 0$, then $\tan (0) - 1 = 0 - 1 = -1 \neq \sec (0) = 1$. Hence, $t = 0$ is **not** a solution of the given equation.

If $t = \pi$, then $\tan (\pi) - 1 = 0 - 1 = -1 = \sec (\pi) = -1$. Therefore, $t = \pi$ **is** a solution of the given equation.

If $t = 2\pi$, then $\tan (2\pi) - 1 = 0 - 1 = -1 \neq \sec (2\pi) = 1$. Hence, $t = 2\pi$ is **not** a solution of the given equation.

The solution set for the given equation, on the given interval, is $\{\pi\}$.

In this chapter, we extended the circular functions to include the addition formulas and the trigonometric functions for multiples of t. The graphs of variations of the basic circular functions were also examined. The chapter concluded with a discussion of the solution of trigonometric equations.

SUPPLEMENTAL EXERCISES

In Exercises 1-10, identify the statement as being true or false and give a reason for each of your answers.

1. The solution of the equation $\sin (t) = \sqrt{3}/2$, on the interval $[0, 2\pi]$, is unique.

2. The product of $\cos (-1.4)$ and $\sec (1.4)$ is -1.

3. The domain of every trigonometric function is $(-\infty, +\infty)$.

4. $\sin (4t) = 2 \sin (2t)$.

5. There are no values of t, on the interval $[\pi/2, 3\pi/2]$, such that $\sin (t) = \cos (t)$.

6. If $f(t) = \sin (t) + \cos (t)$, then $f(t) = \sqrt{2}$ for some values of t in the interval $[0, 2\pi]$.

7. If T is a circular function, then $T(t + 2\pi) = T(t)$ for all real values of t for which the function is defined.

8. On the interval $[\pi/2, 3\pi/2]$, the cosine function is an increasing function.

9. $\tan (-t) = \dfrac{\sin (-t)}{\cos (t)}$.

10. The amplitude of the graph of $y = \sin(2t)$ is 2.

For the graph of each of the functions defined in Exercises 11-15, state the amplitude, period, and phase shift.

11. $y = \sin(-2x)$

12. $y = -3 \cdot \cos(4x)$

13. $y = \sin(x + \pi/8)$

14. $y = -4 \cdot \cos(x - \pi)$

15. $y = (-0.76)\sin(2\pi - 3x)$

In Exercises 16-20, rewrite each of the given expressions in terms of a single circular function.

16. $\sin(3t)\cos(2t) - \cos(3t)\sin(2t)$

17. $\cos^2(t/2) - \sin^2(t/2)$

18. $4 \cdot \sin(3t)\cos(3t)$

19. $\sin(5t)\sin(6t) - \cos(5t)\cos(6t)$

20. $\dfrac{1 - \cos(6t)}{4}$

ANSWERS TO SUPPLEMENTAL EXERCISES

1. False, there are two solutions on the given interval.

2. False, the product is 1.

3. False; only the sine and the cosine functions have the interval $(-\infty, +\infty)$ for their domain.

4. False; $\sin(4t) = 2 \cdot \sin(2t)\cos(2t)$.

5. False; $t = 5\pi/4$ satisfies the equation.

6. True, if $t = \pi/4$.

7. True; 2π is the period of all of the circular functions.

8. False; the cosine function decreases on $[\pi/2, \pi]$ but increases on $[\pi, 3\pi/2]$.

9. True since $\dfrac{\sin(-t)}{\cos(t)} = \dfrac{-\sin(t)}{\cos(t)} = -\tan(t) = \tan(-t)$.

10. False; the amplitude is 1.

11. Amplitude = 1; Period = π; Phase shift is 0 units.

12. Amplitude = 3; Period = $\pi/2$; Phase shift is 0 units.

13. Amplitude = 1; Period = 2π; Phase shift is $\pi/8$ units to the left.

14. Amplitude = 4; Period = 2π; Phase shift is π units to the right.

15. Amplitude = 0.76; Period = $2\pi/3$; Phase shift is $2\pi/3$ units to the right.

16. $\sin(t)$

17. $\cos(t)$

18. $2 \cdot \sin(6t)$

19. $-\cos(11t)$

20. $\dfrac{1}{2} \cdot \sin(3t)$

Index

Index

Notes/Calculations

Notes/Calculations

Notes/Calculations

Notes/Calculations

Notes/Calculations